Natural Prod...
Botanical Medicines
of Iran

Natural Products Chemistry of Global Plants

Series Editor:
Raymond Cooper

This unique book series focuses on the natural products chemistry of botanical medicines from different countries such as Sri Lanka, Cambodia, Brazil, China, Africa, Borneo, Thailand, and Silk Road Countries. These fascinating volumes are written by experts from their respective countries. The series will focus on the pharmacognosy, covering recognized areas rich in folklore as well as botanical medicinal uses as a platform to present the natural products and organic chemistry. Where possible, the authors will link these molecules to pharmacological modes of action. The series intends to trace a route through history from ancient civilizations to the modern day showing the importance to man of natural products in medicines, foods, and a variety of other ways.

RECENT TITLES IN THIS SERIES

Medicinal Plants of Borneo
Simon Gibbons and Stephen P. Teo

Natural Products and Botanical Medicines of Iran
Reza Eddin Owfi

Natural Products of Silk Road Plants
Raymond Cooper and Jeffrey John Deakin

Brazilian Medicinal Plants
Luzia Modolo and Mary Ann Foglio

Medicinal Plants of Bangladesh and West Bengal
Botany, Natural Products, and Ethnopharmacology
Christophe Wiart

Traditional Herbal Remedies of Sri Lanka
Viduranga Y. Waisundara

Natural Products and Botanical Medicines of Iran

Reza E. Owfi

CRC Press
Taylor & Francis Group
Boca Raton London New York

CRC Press is an imprint of the
Taylor & Francis Group, an **informa** business

First edition published 2021
by CRC Press
6000 Broken Sound Parkway NW, Suite 300, Boca Raton, FL 33487-2742

and by CRC Press
2 Park Square, Milton Park, Abingdon, Oxon, OX14 4RN

© 2021 Taylor & Francis Group, LLC

CRC Press is an imprint of Taylor & Francis Group, LLC

Library of Congress Cataloging-in-Publication Data

Names: Owfi, Reza, author.
Title: Natural products and botanical medicines of Iran / Reza Owfi.
Other titles: Natural products chemistry of global plants.
Description: First edition. | Boca Raton : CRC Press, 2020. | Series:
Natural products chemistry of global plants | Includes bibliographical
references and index.
Identifiers: LCCN 2020020218 | ISBN 9780367441739 (paperback) | ISBN
9780367556570 (hardback) | ISBN 9781003008996 (ebook)
Subjects: LCSH: Medicinal plants--Iran. | Materia medica, Vegetable--Iran.
| Herbs--Therapeutic use--Iran. | Botany, Medical--Iran.
Classification: LCC RS180.I7 O94 2020 | DDC 615.3/210955--dc23
LC record available at https://lccn.loc.gov/2020020218

ISBN: 9780367556570 (hbk)
ISBN: 9780367441739 (pbk)
ISBN: 9781003008996 (ebk)

Typeset in Times
by Deanta Global Publishing Services, Chennai, India

Visit the eResources: https://www.routledge.com/9780367441739

Contents

Preface

Natural Products Chemistry of Global Plants

CRC Press is publishing a new book series on the *Natural Products Chemistry of Global Plants*. The first three volumes are in print from Sri Lanka, Bangladesh and West Bengal, to Brazil. This new series will focus on ethno-pharmacology and pharmacognosy, covering recognized areas rich in folklore and botanical medicinal uses as a platform to present the natural products and organic chemistry and where possible, link these molecules to pharmacological modes of action. This book series on the botanical medicines from different countries includes this volume from Iran, and other volumes from Bangladesh, Borneo, Brazil, Cambodia, Cameroon, Ecuador, Madagascar, South Africa, Sri Lanka, the Silk Road, Thailand, Turkey, Uganda, Vietnam, and Yunnan Province (China), and is written by experts from each country. The intention is to provide a platform to bring forward information from regions often under-represented in the information about their folklore.

In this volume, the author, Dr. Reza E. Owfi, has done an outstanding job of describing much of the rich diversity of plants drawing on the culture, folklore, and environment of medicinal plants and natural products of plants in Iran.

Indeed, medicinal plants are an important part of human history, culture, and tradition, and have been used for medicinal purposes for thousands of years. Anecdotal and traditional wisdom concerning the use of botanical compounds is documented in the rich histories of traditional medicines. Many medicinal plants, spices, and perfumes changed the world through their impact on civilization, trade, and conquest. Folk medicine is commonly characterized by the application of simple indigenous remedies. People who use traditional remedies may not understand in Western terms the scientific rationale for why they work but know from personal experience that some plants can be highly effective.

This series provides rich sources of information from each region. An intention of the series of books is to trace a route through history from ancient civilizations to the modern day, showing the important value to humankind of natural products in medicines, foods, and many other ways. Many of the extracts are today associated with important drugs, nutrition products, beverages, perfumes, cosmetics, and pigments, which will be highlighted.

The books will be written for chemistry, biology, and ethnobotany students who are at university level and for scholars wishing to broaden their knowledge in pharmacognosy. Through examples of chosen herbs and plants, the series will describe the key natural products and their extracts with emphasis upon sources, an appreciation of these complex molecules, and applications in science.

In this series, the chemistry and structure of many substances from each region will be presented and explored. Often, books describing folklore medicine do not describe the rich chemistry or the complexity of the natural products and their respective bio-synthetic building blocks. The story becomes more fascinating by drawing upon the chemistry of functional groups to show how they influence the chemical behavior of the building blocks which make up large and complex natural products. Where possible, it will be advantageous to describe the pharmacological nature of these natural products.

R. Cooper Ph.D., Editor-in-Chief
Dept. Applied Biology & Chemical Technology
The Hong Kong Polytechnic University
Hong Kong

1

Introduction

Medicinal plants are one of the most important natural resources which directly have a positive impact on the lives of people. Many of the drugs we use are directly or indirectly obtained from medicinal plant products. The importance of plants has been evident to humans since ancient times as Egyptians had expertise in the use of plants for medical purposes from the 16th century BC. In addition, also from ancient times, peoples in India, China, and Biblical regions reportedly used plants for health benefits. From the 8th to 10th centuries AD, the famous Iranian scientists, Avicenna and Razi, shone in this field and their wisdom continued up to recent times (Zargari, 2014; Jouri & Mahdavi, 2010). In Iran, the most comprehensive investigation of the use of medicinal plants was conducted by Ali Zargari and these results have been published in a five-volume series.

Iran has an area of 1,648,195 km² and is located in the Middle East (shown in Figure 1.1), and is one of the most unique countries in terms of climate. The difference in daily air temperature in the winter, between the warmest and the coldest parts of the country, sometimes reaches more than 50°C. The annual precipitation in the north of the country is sometimes more than 2000 mm and in the Lut Desert, which is one of the warmest places in the world, it drops to less than 50 mm, and the average annual precipitation in Iran is 240 mm (Jafari & Tavili, 2012; Adl, 1970). The following can be found in Iran in terms of altitudes: from −24 m up to +5678 m above sea level; and in terms of temperature, from −30°C up to +50°C. Although 90% of Iran is made up of arid and semi-arid regions, there are generally seven important vegetation types, based on climate, humidity, and soil in the country, and these different areas result in a high diversity of vegetation in Iran:

1. Desert vegetation type with an approximate area of 350,000 km² and annual precipitation less than 100 mm.
2. Steppe vegetation type with an approximate area of 460,000 km² and annual precipitation of 100 mm to 230 mm.
3. Semi-steppe vegetation type with an approximate area of 300,000 km² and annual precipitation of 230 mm to 400 mm
4. Baluchi vegetation type with an approximate area of 80,000 km² and annual precipitation of 100 mm to 200 mm and relative humidity of 60% to 80% due to the proximity to the Sea of Oman and the Persian Gulf.
5. Arid forest vegetation type with an approximate area of 140,000 km² and annual precipitation of 400 mm and more; also with altitudes of 800 m to 2600 m above sea level.

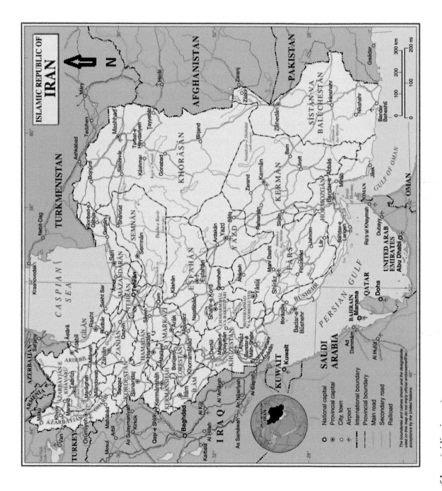

FIGURE 1.1   The location of Iran (vidiani.com).

6. High mountainous vegetation type with an approximate area of 50,000 km^2 and more snowfall than rainfall; also with altitudes of more than 2600 m above sea level.

7. Rain forests of the north (Hyrcanian forests) vegetation type with an approximate area of 20,000 km^2 and annual precipitation of 500 mm to 2000 mm; altitudes from the Caspian Sea coast reach up to 2500 m above sea level, as well as relative humidity more than 80% due to an enclosed region between the Alborz Mountains and the Caspian Sea (Mesdaghi, 2010).

In total, almost 8000 plant species are present in Iran (Hakimi Meybodi, 2009; medplant.Ir) of which more than 2300 species have medicinal, edible, and industrial properties; more than 1700 species are endemic (medplant.Ir).

1.1 Natural Products and Botanical Medicines of Iran

This book provides an overview of the important endemic plants of Iran and their natural products. Their benefits as food, as well as medicinal and industrial usages, are presented. Firstly, it is important to provide a short review of the Boissier plants taxonomy method as this is one of the pioneering methods for classification of medicinal plants in Iran. For clarity of use, all results have been tabulated according to the Boissier plants taxonomy method. The key and detailed information of each species, based on alphabetical family name, are presented in tabular form and summarized in the accompanying tables. After each table, general information about the mentioned plants, based on alphabetical species name, is provided. Illustrations show the plant part, flowers and/or other salient aspects of the plant, followed by key chemical structures and chemical formulas of important substances found in the plant.

Furthermore, in using these tables and writings, it should be noted that most of these plants have various synonym names. On the other hand, some families are given with former names; therefore, their recent names are provided.

With regard to tree species in Iran, most are used for construction, handicrafts, and wood crafts, especially by indigenous people, and many plants are used in the food and cosmetic industries. In addition, "edible" is used to describe plants where some main parts, such as fruit, leaf, root, etc., can be eaten directly without any further processing. Furthermore, most medicinal plants have regional uses by indigenous people, and these are described herein.

1.2 The Boissier Plants Taxonomy Method

This plants taxonomy method was introduced by Pierre Edmond Boissier (1810–1885), who was a Swiss botanist and explorer and amongst the most prolific collectors of the 19th century. He traveled through much of Europe, North Africa, and the Middle East, and produced a vast taxonomic output. This method is one of the pioneering methods for classification of medicinal plants in Iran and it is used by credible related books in the country. Figure 1.2 provides a summary of the Boissier plants taxonomy method.

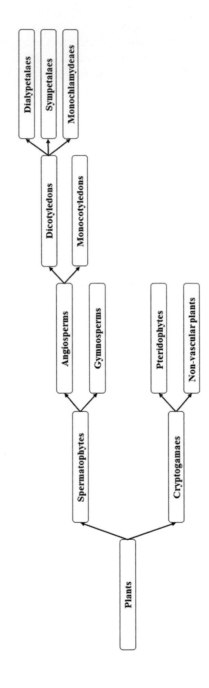

FIGURE 1.2 The summary of the Boissier plants taxonomy method.

Spermatophytes comprise those plants that produce seeds and include the gymnosperms and angiosperms, hence the alternative name is seed plants. They have flowers.

Angiosperms are the most diverse group of land plants. Like gymnosperms, angiosperms are seed-producing plants. However, they are distinguished from gymnosperms by characteristics including flowers, endosperm within the seeds, and the production of fruits that contain the seeds. The ovules and seeds are enclosed in an ovary, form the embryo and endosperm by double fertilization, and typically each flower is surrounded by a perianth composed of two sets of floral envelopes comprising the calyx and corolla.

Gymnosperms (pinophyta) are a group of seed-producing plants. The name is based on the unenclosed condition of their seeds (called ovules in their unfertilized state). The gymnosperms and angiosperms together compose the spermatophytes or seed plants.

Dicotyledons (dicots) have one of the typical characteristics in that the seed has two embryonic leaves or cotyledons. They produce an embryo with two cotyledons and usually have floral organs arranged in cycles of four or five and leaves with reticulate venation.

Monocotyledons (monocots) are flowering plants (angiosperms), the seeds of which typically contain only one embryonic leaf, or cotyledon. They have an embryo with a single cotyledon, usually parallel-veined leaves, and floral organs arranged in cycles of three.

Dialypetalaes (choripetalaes) are plants with the floral corolla is divided into distinct petals.

Sympetalaes (metachlamydeaes, gamopetales) is a group in which the flowers have a separate calyx and corolla with the petals fused and at least at the base of the corolla, a condition known as sympetaly. In fact, the petals of the flowers are united.

Monochlamydeaes (apetales) is a group which have plants with flowers that have either a calyx or corolla, but not both. In other words, flowers that lack petals or sepals but not both.

Cryptogamaes are plants (in the widest sense of the word) that reproduce by spores, without flowers or seeds. In fact, they are plants or plant-like organisms (such as a fern, moss, alga, or fungus) which reproduce by spores and not by producing flowers or seeds.

Pteridophytes are vascular plants (with xylem and phloem) that disperse spores, because pteridophytes produce neither flowers nor seeds. In fact, they are vascular plants (such as a fern) that have roots, stems, and leaves, but lack flowers or seeds.

Vascular plants (tracheophytes) are a large group of plants that are defined as those land plants that have lignified tissues (xylem) for conducting water and minerals throughout the plant. They also have a specialized non-lignified tissue (phloem) to conduct products of photosynthesis.

Non-vascular plants are plants without a vascular system consisting of the xylem and phloem. This group is equal to thallophytes in other plants taxonomy methods. Although non-vascular plants lack these particular tissues, many possess simpler tissues that are specialized for the internal transport of water (Zargari, 2014).

2

Spermatophytes: Angiosperms, Dicotyledons, Dialypetalaes

Spermatophytes are split into several sections. This chapter reviews angiosperms, dicotyledons, and dialypetalaes which are listed in Table 2.1. A total of 113 species is provided and described in detail below. As each species is presented, information on the taxonomy, plant use, and specifically the key natural products isolated from each plant is provided.

In this group, the petals are separate, but in several of them the petals may not exist or are even continuous (Azadbakht, 2000).

Due to the adaptation of this group to the climatic conditions of Iran and their distribution and diversity in the country, most of the medicinal plants in Iran belong to this group. It consists of important families: cruciferae (brassicaceae), leguminosae (papilionaceae), ranunculaceae, rosaceae, rutaceae, and umbelliferae (apiaceae), all of which have important uses.

Acacia farnesiana (Figure 2.1) is a fast-growing deciduous bushy tree that reaches up to 9 m in size. The species is hermaphrodite (has both male and female organs). It can fix nitrogen and has a It has a preference for light (sandy), medium (loamy), heavy (clay), and well-drained soil, and can grow in nutritionally poor soil. It can also grow in very acid, very alkaline, and saline soils. It cannot grow in the shade. It prefers dry or moist soil and can tolerate drought. The bark is astringent and demulcent. Along with the leaves and roots, it is used for medicinal purposes.

A bark decoction of "Colombians bathe" (sweet acacia) is known as a treatment for typhoid. The gummy roots are chewed as a treatment for sore throats and a decoction of the gum from the trunk has been used in the treatment of diarrhea. An infusion of the flowers has been used as a stomachic. It is also used in the treatment of dyspepsia and neuroses. The flowers are added to ointment, which is rubbed on the forehead to treat headaches (Zargari, 2014; Mozaffarian, 2011; pfaf.org; Claudia *et al.*, 2018).

Albizia lebbek (Figure 2.2) is a fast-growing deciduous tree that reaches up to 15 m in size. It prefers light (sandy), medium (loamy), heavy (clay), and well-drained soil, and can grow in nutritionally poor soil. It can also grow in very acidic, very alkaline, and saline soils. It cannot grow in the shade. It prefers moist soil. The plant is not wind tolerant.

The leaves and seeds are used in the treatment of eye problems such as ophthalmia. The bark is astringent; it is taken internally to treat diarrhea, dysentery, and piles. The bark is used externally to treat boils. The flowers are applied locally to maturate boils and alleviate skin eruptions. The powdered seeds are used to treat scrofula. Saponins (e.g. $C_{55}H_{86}O_{24}$) from the pods and roots have spermicidal properties. The bark is used as a fish poison. Rutin ($C_{27}H_{30}O_{16}$), a red dye obtained from the bark, causes skin irritation. The pods containing saponins are not eaten in large amounts by sheep, although cattle eat them readily (Zargari, 2014; Mozaffarian, 2011; pfaf.org; Ma *et al.*, 1997; Vinod *et al.*, 2010).

TABLE 2.1

Spermatophytes: Angiosperms, Dicotyledons, Dialypetalaes

Family Name	Species Name	Major Features of Botany	Usable Parts	Some Important Substances	Edible, Some Industrial and Medicinal Benefits
Anacardiaceae	*Mangifera indica*	Evergreen Tree—Aromatic Fruit—Small and White Blossom	Flower—Fruit—Leaf—Bark—Seed	Astragalin—Fisetin—Gallic Acid—Fiber—Mangiferin—Quercetin—Methyl Gallate—Gallothannin—Vitamins A and C	Edible—Astringent—Appetizer—Dialysis—Anti-Diuretic, Anti-Diarrheal, and Anti-Emetic—Treating Blood Pressure, Angina, Asthma, Coughs, Diabetes, Colds, Dysentery, Scorpion Stings, Hemorrhage, and Stomach Pain
Anacardiaceae	*Pistacia vera*	Tree—Leaf without Trichome—Dry Fruit	Fruit—Seed—Gum	Lutein—Carotenoids—Phytosterols—Vitamin A	Edible—Tonic—Sedative—Treating Anemia—Antioxidant, Anti-Microbial, and Anti-Mutagenic
Anacardiaceae	*Rhus coriaria*	Bushy Tree—Small Fruit—Leaves Change to Red and Green in Autumn	Fruit—Leaf—Seed	Tannic Acid	Edible—Anti-Diarrheal—Coagulator—Astringent—Diuretic—Styptic—Tonic—Treating Dysentery, Hemoptysis, and Conjunctivitis
Araliaceae	*Hedera helix*	Evergreen—Climbing or Ground—Creeping—Green and Leather Leaf	Root—Bud—Leaf	Hederin—Hederic Acid—Chlorogenic Acid—Emetine—Triterpenes	Antiseptic, Anti-Bacterial, and Anti-Spasmodic—Treating Rheumatic—Astringent, Cathartic, Diaphoretic, Stimulant, Sudorific, Vasoconstrictor, Vasodilator—Vermifuge—Aperient—Emetic—Emmenagogue
Berberidaceae	*Berberis vulgaris*	Spiny Bushy Tree—Yellow Flower and Wood	Root—Flower—Fruit—Bark of Stem—Bark of Root	Berberine—Berbamine—Palmatine	Edible—Anti-Diarrheal, Anti-Bleeding, and Anti-Gingivitis
Cornaceae	*Cornus mas*	Bushy Tree—Yellow and Small Flower	Fruit—Flower—Bark	Glyoxylic Acid—Tannins	Edible—Anti-Fever and Anti-Diarrheal—Tonic—Astringent

(Continued)

TABLE 2.1 (CONTINUED)

Spermatophytes: Angiosperms, Dicotyledons, Dialypetalaes

Family Name	Species Name	Major Features of Botany	Usable Parts	Some Important Substances	Edible, Some Industrial and Medicinal Benefits
Cornaceae	*Cornus sanguinea*	Bushy Tree—Leaves Change to Red in Summer	Fruit—Bark of Stem—Leaf	Quercetin	Anti-Fever—Emetic—Astringent—Febrifuge
Cruciferae (Brassicaceae)	*Brassica napus*	Herbal—Leaves without Trichome—Swollen Root	Root	Vitamins A, B₉, C, and K—Fiber—Lutein	Industrial (Animal and Bird Food Making)—Edible—Anti-Scurvy—Diuretic—Treating Kidney Stones
Cruciferae (Brassicaceae)	*Brassica nigra*	Herbal—Yellow Flower	Seed	Moronic Acid—Sinigrin—Sinapine	Industrial (Soap Making)—Stomach Tonic—Anti-Constipation—Treating Rheumatism, Epilepsy, and Toothache
Cruciferae (Brassicaceae)	*Brassica oleracea*	Grass—Leaf with Trichome	Leaf	Sulforaphane—Vitamin C	Edible—Anti-Scurvy—Wound Healing—Treating Asthma and Arthritis
Cruciferae (Brassicaceae)	*Cheiranthus cheiri*	Herbal—Evergreen—Narrow Leaves with lots of Brown Hairs—Yellow Flower	Seed—Leaf	Quercetin—Cheiranthin—Quinoline	Treating Headache and Stuffy Nose—Diuretic—Laxative—Anti-Rheumatic and Anti-Spasmodic
Cruciferae (Brassicaceae)	*Descurainia sophia*	Herbal—Stem Branches at Ground Level	Seed—Leaf	Oleic Acid—Linoleic Acid—Palmitic Acid—Linolenic Acid—Benzyl	Edible—Wound Healing—Anti-Fever, Anti-Worm, and Anti-Scorbutic—Astringent—Diuretic—Expectorant—Febrifuge—Laxative—Restorative—Tonic
Cruciferae (Brassicaceae)	*Lepidium latifolium*	Herbal—Big Leaf—Small Flower	Leaf—Flower	Lepidine	Edible—Tonic—Anti-Scurvy—Dialysis—Diuretic—Depurative—Stomachic

(Continued)

TABLE 2.1 (CONTINUED)

Spermatophytes: Angiosperms, Dicotyledons, Dialypetalaes

Family Name	Species Name	Major Features of Botany	Usable Parts	Some Important Substances	Edible, Some Industrial and Medicinal Benefits
Cruciferae (Brassicaceae)	*Lepidium sativum*	Light Green Leaf and Stem—Pink or White Flower	Leaf—Stem—Seed—Root	Carotene—Vitamin C	Edible—Appetizer—Dialysis—Anti-Scurvy—Diuretic—Stimulant—Galactogogue—Treating Secondary Syphilis and Tenesmus
Cruciferae (Brassicaceae)	*Nasturtium officinale*	Herbal—Fleshy Leaves—Growing in Clear Water	Leaf—Juice	Gluconasturtiin—Phenyl Ethyl—Vitamin C	Edible—Mucus Producing—Treating Gout and Rheumatism—Depurative—Diuretic—Expectorant—Purgative—Hypoglycemic—Odontalgic—Stimulant—Stomachic—Anti-Scurvy
Cruciferae (Brassicaceae)	*Raphnus sativus*	Herbal—Swollen Root—White, Yellow, or Violet Flower	Juice of Leaf—Root—Leaf—Seed	Sulforaphane—Raphanol	Edible—Treating Whooping Cough, Gall Bladder Stone, Asthma, and Rheumatism—Stimulates Appetite and Digestion—Tonic—Laxative—Astringent—Digestive—Diuretic—Anti-Scorbutic, Anti-Tumor, and Anti-Spasmodic
Cruciferae (Brassicaceae)	*Sinapis alba*	Herbal—Covered with Trichome—Yellow Flower	Seed—Leaf—Flower—Seedling Plant	Sinalbin—Vitamins A and C	Edible—Diuretic—Laxative—Appetizer—Carminative—Diaphoretic—Digestive—Emetic—Expectorant—Rubefacient—Stimulant—Treating Coughs and Tuberculosis—Anti-Bacterial and Anti-Fungal
Cruciferae (Brassicaceae)	*Sinapis arvensis*	Herbal—Very Small and Black Seed	Seed—Leaf—Flower	Sinalbin—Gibberellic Acid	Edible—Diuretic—Laxative—Appetizer—Treating Depression

(Continued)

TABLE 2.1 (CONTINUED)
Spermatophytes: Angiosperms, Dicotyledons, Dialypetalaes

Family Name	Species Name	Major Features of Botany	Usable Parts	Some Important Substances	Edible, Some Industrial and Medicinal Benefits
Cucurbitaceae	*Citrullus colocynthis*	Herbal—Stem Covered by Trichome—Leaf with Trichome	Fruit	Citrulline—Colocynthin—Colocynthein—Citrullol	Edible—Strong Aperient—Treating Visceral Paralysis
Cucurbitaceae	*Citrullus vulgaris*	Shrub—Yellow Flower—Fleshy Fruit	Fruit—Seed	Citrulline—Vitamins A and C—Carotene—Lycopene	Edible—Diuretic—Anti-Fever and Anti-Infection
Cucurbitaceae	*Cucumis melo*	Shrub—Yellow Flower—Fleshy Fruit	Fruit—Seed	Vitamins A and C—Cellulose	Edible—Anti-Worm, Anti-Tussive—Diuretic—Digestive—Febrifuge—Vermifuge
Cucurbitaceae	*Cucumis sativus*	Shrub—Stem with Rough Hairs—Big Yellow Flower	Fruit—Seed	Mucilage—Erepsin	Edible—Dialysis—Diuretic—Anti-Worm—Skin Hygiene—Cooling—Tonic—Vermifuge
Fumariaceae	*Fumaria parviflora*	Herbal—Dusty Leaves—White Flower	Whole Plant	Protopine—Fumoficinaline	Sweaty—Appetizer—Dialysis—Tonic
Geraniaceae	*Geranium robertianum*	Herbal—Multi Parts with Funky Leaves—	Whole Plant	Ellagic Acid—Geranine	Diuretic—Tonic—Astringent—Anti-Rheumatic
Hypericaceae (Cluciaceae)	*Hypericum perforatum*	Herbal—Stem with Two Longitudinal Lines—Oval Leaves	Twig with Flower	Hyperin—Hypericin—Pectin—Pseudohypericin—β-Sitosterol—Choline	Tonic—Digestive—Painkiller—Anti-Bile—Treating HIV, Wounds, Sores, Ulcers, Swellings, and Rheumatism
Leguminosae (Papilionaceae)	*Acacia farnesiana*	Bushy Tree—Spiny Twig—Yellow Flower	Root—Flower—Bark	Polyphenol—Cetane—Anisic Aldehyde—Kaempferol	Anti-Worm—Anti-Nervous Pain—Astringent
Leguminosae (Papilionaceae)	*Albizia lebbek*	Tree without Thorn—Flower in White Pink and Red	Stem—Bark—Root—Pod	Saponins—Rutin	Anti-Worm—Treating Itching—Treating Asthma

(Continued)

TABLE 2.1 (CONTINUED)

Spermatophytes: Angiosperms, Dicotyledons, Dialypetalaes

Family Name	Species Name	Major Features of Botany	Usable Parts	Some Important Substances	Edible, Some Industrial and Medicinal Benefits
Leguminosae (Papilionaceae)	*Alhagi camelorum*	Herbal—Barbed Stem	Manna	Melezitose	Laxative Medication— Sweetener
Leguminosae (Papilionaceae)	*Arachis hypogaea*	Herbal—Yellow Flower	Fruit	Oleic Acid—Palmitic Acid— Aflatoxin B_1, B_2, G_1, and G_2	Industrial (Camphor Oil and Insecticide Making)—Edible—Aperient— Demulcent—Emollient—Pectoral— Anti-Inflammatory—Aphrodisiac— Decoagulant
Leguminosae (Papilionaceae)	*Astragalus sp.*	Herbal, Shrub or Bushy Tree—Spiny or without Spin—Pink, Red, Blue, Yellow, or Violet Flower	Sarcocolla— Manna—Tragacanth	Tragacanthin—Uronic Acid—Saponins—Bassorin— Cycloastragenol	Industrial (Tragacanth for Mucilage Making)—Sarcocolla is a Painkiller and Coagulator
Leguminosae (Papilionaceae)	*Cassia fistula*	Yellow Leaf—Yellow Flower	Brain of Fruit	Oxy Methyl—Anthraquinone— Glycoside	Laxative—Aperient—Anti- Inflammatory
Leguminosae (Papilionaceae)	*Cercis siliquastrum*	Beautiful Bushy or Small Tree—Purple Flower	Flower—Bud— Seedpod	Oleic Acid—Palmitic Acid—Anthocyanin	Edible—Astringent—Treating Bronchitis—Antioxidant, Anti- Inflammatory, Anti-Coagulant, Anti-Diabetic, Anti-Malarial, and Anti-Microbial
Leguminosae (Papilionaceae)	*Cicer arietinum*	Herbal—White Flower	Seed	Lecithin—Galactan—Lysine— Linoleic Acid	Edible—Anti-Worm—Emmenagogue— Diuretic—Treating Diarrhea, Diabetes, and Anxiety
Leguminosae (Papilionaceae)	*Ceratonia siliqua*	Tree—Knotted Twig	Bark—Seed—Leaf— Fruit	Galactomannan—Gallic Acid—Phytin	Industrial (Mucilage Making)—Anti- Diarrheal—Treating Tuberculosis—Appetizer—Diuretic

(Continued)

TABLE 2.1 (CONTINUED)

Spermatophytes: Angiosperms, Dicotyledons, Dialypetalaes

Family Name	Species Name	Major Features of Botany	Usable Parts	Some Important Substances	Edible, Some Industrial and Medicinal Benefits
Leguminosae (Papilionaceae)	*Faba vulgaris*	Herbal—Big and White Flower	Flower—Fruit—Seed	Proteoses—Vicilin	Edible—Tonic—Treating Fatigue and Paroxysm
Leguminosae (Papilionaceae)	*Glycyrrhiza glabra*	Herbal—Yellow and Violet Flower	Stolon—Root	Glycyrrhizin	Treating Coughs, Bronchitis, Gastritis, and Inflammation of Stomach—Demulcent—Diuretic—Emollient—Expectorant—Laxative—Moderately Pectoral—Tonic
Leguminosae (Papilionaceae)	*Lens culinaris*	Herbal—White and Small Flower	Seed	Starch—Vitamin B$_9$	Edible—Treating Inflammation, Constipation—Laxative
Leguminosae (Papilionaceae)	*Medicago sativa*	Herbal—Big Blue and Violet Flower	Whole Plant	Asparagine—Arginine—Vitamin K—Carotene—Chlorophyll—Tricin—Saponins—Canavanine	Treating Rickets, Asthma, and Diabetes—Anti-Scorbutic- Anti-Aperient—Diuretic—Oxytocic—Hemostatic—Nutritive—Stimulant—Tonic—Febrifuge
Leguminosae (Papilionaceae)	*Melilotus officinalis*	Herbal—Yellow, Small, and Aromatic Flower—Without Trichome Fruit	Twig with Flower—Leaf	Coumarin—Melilotic Acid	Treating Chest Pain, Eye Inflammation, Rheumatic Pain, and Swollen Joints—Anti-Spasmodic—Carminative—Diuretic—Emollient—Expectorant—Sedative—Vulnerary—Diuretic
Leguminosae (Papilionaceae)	*Onobrychis viciifolia*	Herbal—Pink to White Flower	Leaf	Flavonols—Glycosides—Tannins—Polyphenols	Sweaty—Appetizer
Leguminosae (Papilionaceae)	*Phaseolus vulgaris*	Herbal—Flower in White and Yellow	Green Pod—Root—Seed	Arginine	Edible—Diuretic—Hypoglycemic—Hypotensive—Treating Rheumatism, Arthritis, Diabetes, and Cancer of Blood

(Continued)

TABLE 2.1 (CONTINUED)

Spermatophytes: Angiosperms, Dicotyledons, Dialypetalaes

Family Name	Species Name	Major Features of Botany	Usable Parts	Some Important Substances	Edible, Some Industrial and Medicinal Benefits
Leguminosae (Papilionaceae)	*Pisum sativum*	Herbal—Very Big Stipule	Seed	Vicilin—Legumin	Edible—Diuretic—Contraceptive—Fungistatic—Spermicidal
Leguminosae (Papilionaceae)	*Prosopis spicigera* (*P. cineraria*)	Evergreen—Spiny Bushy Tree—Yellow Flower	Flower—Pod—Manna—Bark	Piperidine—Fatty Acids—Flavanones	Industrial (Muscovite Gum Making)—Tonic—Astringent—Demulcent—Pectoral—Anthelmintic—Refrigerant—Anti-Worm—Treating Asthma and Bronchitis
Leguminosae (Papilionaceae)	*Robinia pseudo-acacia*	Tree—Spiny—Gray Bark—White Flower	Root—Flower—Bark	Flavonoids—Robinin—Kaempferol—Benzaldehyde	Tonic—Laxative—Astringent—Anti-Spasmodic—Anti-Diuretic—Emollient—Emetic—Purgative—Cholagogue
Leguminosae (Papilionaceae)	*Trigonella foenum graecum*	Herbal—Single and Yellow Flower	Stem—Seed	Trigonelline—Vitamin B_3—Saponins	Edible—Tonic—Softener—Carminative—Demulcent—Deobstruent—Emollient—Cardiotonic—Diuretic—Expectorant—Febrifuge—Galactogogue—Hypoglycemic—Laxative—Parasiticide—Restorative—Anti-Cholesterolemic, Anti-Phlogistic, Anti-Inflammatory, and Anti-Tumor
Leguminosae (Papilionaceae)	*Vicia sativa*	Herbal—Violet Flower	Seed	Vicine—Legumin	Edible—Treating Measles, Smallpox, and Inflammation
Linaceae	*Linum usitatissimum*	Herbal—Without Trichome—Narrow Leaf	Seed—Flower	Vitamin C—Linoleic Acid—Linamarin—Glutamic Acid	Industrial (Weaving)—Softener—Treating Diabetes, Chronic Constipation, and Coughs—Diuretic—Cardiotonic

(Continued)

TABLE 2.1 (CONTINUED)

Spermatophytes: Angiosperms, Dicotyledons, Dialypetalaes

Family Name	Species Name	Major Features of Botany	Usable Parts	Some Important Substances	Edible, Some Industrial and Medicinal Benefits
Lythraceae	*Lawsonia inermis*	Bushy Tree—Spiny Twig with Grey Bark	Leaf	Tannins—Fatty Substances—Lowson (Dye)	Industrial (Hair, Body, and Fiber Color Making)—Anti-Sweat, Anti-Microbial, and Anti-Fungal—Treating Cancer and Eczema
Magnoliaceae	*Chimonanthus fragrans (Ch. praecox)*	Bushy Tree—Flower in Lemon—Jar-Like Fruit	Root—Leaf—Flower	Calycanthine—O-cymene—Terpinene—Eucalyptol—Linalool—Benzyl Alcohol—Benzyl Acetate—Terpineol—Indole	Stomach Tonic—Treating Rheumatism, Hemorrhages, Strains, Cuts, and Colds
Magnoliaceae	*Illicium verum*	Evergreen Tree—Thick Branch—Grey Bark	Fruit	Anisic Acid—Shikimic Acid—Anethole	Stomach Tonic—Treating Bloating—Mucus Producing—Carminative—Diuretic—Odontalgic—Stimulant
Malvaceae	*Althaea officinalis*	Herbal—Stem with Trichome—Long, Spindle-Like Root	Root—Flower—Leaf	Mucilage—Flavonoids—Tannins—Phenolic Acids	Softener—Treating Bronchitis, Coughs, Skin, Mouth, and Throat Irritation
Malvaceae	*Gossypium herbaceum*	Shrub—Leaf with Trichome—Light Yellow Flower	Cotton Fibers—Leaf—Seed—Bark of Root	Choline—Sitosterol—Yellow and Red Gossypol	Industrial (Weaving)—Anti-Fever—Emmenagogue—Treating Dysentery, Intermittent, and Fibroids
Myrtaceae	*Eucalyptus globulus*	Tree—Firm Wood	Leaf	Eucalyptol—Tannins—Citronellal	Disinfectant—Expectorant—Febrifuge—Hypoglycemic—Stimulant—Astringent—Anti-Fever
Myrtaceae	*Myrtus communis*	Evergreen—Bushy Tree—Leather Leaf—Big and White Flower	Leaf—Fruit—Flower	Myrtidana—Tannins—Linolein—Myrtol	Disinfectant—Astringent—Stomach Tonic—Carminative—Treating Acne, Wounds, Gum infections, Gingivitis, and Hemorrhoids

(Continued)

TABLE 2.1 (CONTINUED)

Spermatophytes: Angiosperms, Dicotyledons, Dialypetalaes

Family Name	Species Name	Major Features of Botany	Usable Parts	Some Important Substances	Edible, Some Industrial and Medicinal Benefits
Oxalidaceae	*Oxalis corniculata*	Herbal—Stem with Trichome	Whole Plant	Oxalic Acid—Carotene—Vitamin B_3 and C	Edible—Appetizer—Emmenagogue—Diuretic—Anthelmintic—Anti-Phlogistic—Astringent—Depurative—Febrifuge—Lithontripic—Stomachic—Styptic
Papaveraceae	*Papaver rhoeas*	Herbal—Leaf with Trichome—Single and Red Flower	Flower—Petal—Seedpod—Leaf—Seed	Papaverine—Morphine—Codeine—Thebaine—Narceine	Laxative—Treating Coughs, Jaundice, Gout, Cancer, and Insomnia
Papaveraceae	*Papaver somniferum*	Herbal—Stem and Leaf with Trichome—White, Mauve, or Red Flower	Whole Plant Especially Sepal and Seed—Latex of Seedpod	Morphine—Thebaine—Papaverine—Noscapine—Codeine—Oripavine	Edible—Treating Headache and Runny Nose—Reduction of Libido—Anti-Tussive—Sedative—Astringent—Hypnotic—Emmenagogue
Portulacaceae	*Portulaca oleracea*	Herbal—Fleshy	Leaf—Stem—Seed	Glutathione—Melatonin—Carotene—Omega-3	Edible—Tonic—Depurative—Diuretic—Febrifuge—Vermifuge—Anti-Fever, Anti-Scurvy, and Anti-Bacterial—Treating Stomachache, Earache, and Headache
Punicaceae (Granataceae) (recently, Lythraceae)	*Punica granatum*	Bushy Tree—Rugged and Firm Stem—Multiple Spiny Twig	Leaf—Flower—Root—Fruit—Bark of Root—Bark of Stem—Seed	Vitamin B_9, C, and K—Fiber—Pelletierine—Tannins	Edible—Treating Loss of Appetite, Diarrhea, Dysentery, Stomachache, Coughs, Hemorrhage, and Anemia—Astringent—Refrigerant—Emmenagogue—Laxative

(Continued)

TABLE 2.1 (CONTINUED)

Spermatophytes: Angiosperms, Dicotyledons, Dialypetalaes

Family Name	Species Name	Major Features of Botany	Usable Parts	Some Important Substances	Edible, Some Industrial and Medicinal Benefits
Ranunculaceae	*Aquilegia vulgaris*	Herbal—Dusty Upper Surface of the Leaves—Blue and Purple Flower	Fruit with Seed—Leaf	Flavonoids—Alkaloids	Sweaty—Diuretic—Anti-Scurvy
Ranunculaceae	*Consolida regalis*	Herbal—Beautiful Blue and White Flower	Seed	Delsoline—Kaempferol	Strong Aperient—Skin Irritating
Ranunculaceae	*Nigella sativa*	Herbal—Multi-Part Leaves—Flower in White	Seed	Nigellone	Aperient—Emmenagogue—Anthelmintic—Carminative—Diaphoretic—Digestive—Diuretic—Emmenagogue—Galactogogue—Laxative—Stimulant—Treating Stomach Pains and Spasms
Ranunculaceae	*Nymphaea alba*	On the Surface of Water—Flower Grows On or Above Water	Rhizome—Root—Flower	Nupharin—Aporphine	Tonic—Anodyne—Astringent—Cardiotonic—Demulcent—Sedative—Anti-Scrofulatic
Ranunculaceae	*Paeonia corallina (P. mascula)*	Herbal—Underground Stem	Root—Petal	Polyphenols—Catechin—Theanine—Caffeine	Treating Coughs, Hemorrhoids, and Varicose Veins—Painkiller—Tonic—Anti-Spasmodic
Ranunculaceae	*Ranunculus sceleratus*	Herbal—Yellow Flower—Hollow Stem	Seed—Leaf—Root	Anemonin—Serotonin—Protoanemonin	Treating Infested Wound, Rheumatism, Spermatorrhoea, and Scabies—Anodyne—Diaphoretic—Tonic—Emmenagogue—Rubefacient—Anti-Spasmodic

(Continued)

TABLE 2.1 (CONTINUED)

Spermatophytes: Angiosperms, Dicotyledons, Dialypetalaes

Family Name	Species Name	Major Features of Botany	Usable Parts	Some Important Substances	Edible, Some Industrial and Medicinal Benefits
Rhamnaceae	*Frangula alnus*	Shrub—Flat and Dark Bark of Stem	Twig—Bark of Stem	Frangulin—Anthraquinone—Anthrone—Anthranol	Anti-Obesity—Aperient—Cathartic—Cholagogue—Laxative—Tonic—Vermifuge
Rhamnaceae	*Ziziphus jujuba*	Tree—Small, Yellow and Green Flower	Fruit—Seed—Root	Triterpenoids—Vitamins A, B_1, B_2, B_3, B_6, and C—Alkaloids—Saponins—Flavonoids	Edible—Laxative—Anodyne—Pectoral—Refrigerant—Hypnotic—Narcotic Sedative—Stomachic—Styptic—Tonic—Treating Diarrhea, Pharyngitis, Bronchitis, Anemia, Irritability, Strangury, and Hysteria—Anti-Cancer
Rosaceae	*Amygdalus communis*	Tree—Plain and Oval Leaf—Pink Flower	Flower—Bark—Fruit—Seed Kernel	Aspartic Acid—Amygdalin—Glycine—Arginine—Choline—Glutamic Acid—Taxifolin—Vitamins B_2, B_3, and E	Edible—Anti-Bile—Diuretic—Treating Cough—Aperient
Rosaceae	*Armeniaca vulgaris* (*Prunus armeniaca*)	Tree—Big, White and Pink Flower	Seed Kernel—Fruit	Conglutin—Pangamic Acid—Amygdalin—Citric Acid—Tartaric Acid	Edible—Astringent—Dialysis—Treating Fevers, Coughs, and Colds
Rosaceae	*Cerasus avium* (*Prunus avium*)	Long-Living Tree—Aspheric Fruit	Fruit	Vitamin B_9—Choline—Amygdalin—Prunasin	Edible—Skin Hygiene—Treating Kidney, Cystitis, Edema, Bronchial Complaints, and Anemia—Astringent—Diuretic—Tonic

(Continued)

TABLE 2.1 (CONTINUED)

Spermatophytes: Angiosperms, Dicotyledons, Dialypetalaes

Family Name	Species Name	Major Features of Botany	Usable Parts	Some Important Substances	Edible, Some Industrial and Medicinal Benefits
Rosaceae	*Cerasus vulgaris* (*Prunus cerasus*)	Tree—With Lots of Shoots—Aspheric Fruit	Fruit	Vitamin B_9—Choline—Amygdalin—Prunasin	Edible—Treating Kidney and Liver Diseases
Rosaceae	*Cydonia oblonga*	Tree—Dark Stem—Leaf with Trichome—Big Flower in Red or Pink	Fruit—Seed—Leaf	Mucilage—Pectin—Amygdalin—Malic Acid—Tannins	Edible—Stomach Tonic—Anti-Diarrheal—Painkiller—Treating Insomnia
Rosaceae	*Fragaria vesca*	Herbal—White Flower	Leaf—Rhizome—Fruit	Tannins—Glucosides—Salicylic Acid	Edible—Treating Shortness of Breath, Body Hypothermia, and Diarrhea—Dialysis
Rosaceae	*Malus orientalis*	Tree—Upper Leaf Level with Trichome—Big,White and Pink Flower	Leaf—Fruit—Seed—Bark of Stem—Bark of Root	Phlorizin—Phloretin	Edible—Diuretic—Astringent—Anthelmintic—Refrigerant—Soporific—Anti-Fever—Treating Intermittent, Remittent, and Bilious Fevers
Rosaceae	*Mespilus germanica*	Spiny Tree—Big Flower in White and Pink	Fruit—Leaf—Seed	Malic Acid—Tannins—Citric Acid—Tartaric Acid—Quinine	Edible—Astringent—Laxative—Lithontripic—Treating Hemorrhage
Rosaceae	*Persica vulgaris* (*Prunus persica*)	Tree—Narrow and Long Leaf	Bark—Fruit—Leaf—Flower	Amygdalin—Conglutin	Edible—Aperient—Skin Hygiene—Astringent—Demulcent—Diuretic—Expectorant—Febrifuge—Laxative—Parasiticide—Sedative—Emollient, Hemolytic, Anti-Asthmatic, Anti-Tussive, and Anti-Fever
Rosaceae	*Prunus domestica*	Tree—Oval Leaf—Big Fruit	Fruit—Bark	Vitamin C—Pyranose—Amygdalin—Prunasin	Edible—Tonic—Laxative—Stomachic—Febrifuge—Treating Gout and Rheumatism

(Continued)

TABLE 2.1 (CONTINUED)

Spermatophytes: Angiosperms, Dicotyledons, Dialypetalaes

Family Name	Species Name	Major Features of Botany	Usable Parts	Some Important Substances	Edible, Some Industrial and Medicinal Benefits
Rosaceae	*Prunus spinosa*	Spiny Bushy Tree—White Flower	Bark—Stem—Leaf—Flower—Fruit	Kaempferol—Phlobaphenes—Amygdalin—Prunasin	Edible—Astringent—Diuretic—Laxative—Dialysis—Aperient—Astringent—Depurative—Diaphoretic—Febrifuge—Stomachic—Anti-Fever
Rosaceae	*Pyrus communis*	Tree—White and Pink Flower	Stem—Fruit—Leaf—Bark—Bud—Flower	Arbutin—Phlorizin—Tannins—Astragalin—Sorbitol—Vitamin C	Edible—Astringent—Diuretic—Sedative—Febrifuge—Treating Psychosis and Wounds—Antioxidant and Anti-Inflammatory
Rosaceae	*Rosa canina*	Bushy Tree—Flower in White and Pink	Petal—Hip—Gall—Seed—Flower—Fruit	Tannins—Malic Acid—Citric Acid—Flavonoids—Vitamins A, C, and E	Treating Bloody Diarrhea and Kidney Inflammation—Anti-Scurvy—Astringent—Carminative—Diuretic—Laxative—Ophthalmic—Tonic—Vermifuge
Rosaceae	*Rosa damascena*	Bushy Tree—Pink Petal	Petal—Fruit—Bud	Stearoptene—Oleoptene—Farnesol—Geraniol—Nerol—Linalool—Citronellol—Flavonoids—Vitamins A, C, and E—Fatty Acids	Laxative—Astringent—Aperient—Cardiac—Tonic
Rosaceae	*Rosa gallica*	Spiny Bushy Tree—Red and Aromatic Flower	Bud—Fruit—Flower—Petal	Cyanine—Vitamins A, C, and E—Flavonoids—Fatty Acids	Astringent—Treating Hemorrhage—Skin Hygiene

(Continued)

TABLE 2.1 (CONTINUED)

Spermatophytes: Angiosperms, Dicotyledons, Dialypetalaes

Family Name	Species Name	Major Features of Botany	Usable Parts	Some Important Substances	Edible, Some Industrial and Medicinal Benefits
Rosaceae	*Rubus caesius*	Bushy Tree—Blue Fruit—Spiny Twig	Leaf—Bud—Stem—Fruit—Bark of Root	Albuminoids—Organic Acids—Tannins—Methanol—Ethanol—Vitamins A and C	Edible—Anti-Diarrheal, Anti-Inflammatory, Anti-Viral, Anti-Microbial, and Anti-Tumor—Astringent—Tonic—Dialysis—Coagulator—Appetizer—Treating Stomach Problems
Rosaceae	*Sanguisorba minor*	Herbal—With Rhizome—Yellow to Brown Flower	Root—Leaf	Tannins—Minerals	Coagulator—Diuretic—Astringent—Appetizer—Diaphoretic—Styptic—Treating Gout, Rheumatism, and Eczema
Rosaceae	*Sorbus torminalis*	Tree—Dark Grey Bark of Trunk	Fruit—Leaf	Tannins—Flavonols—Isoquercitrin—Chlorogenic Acid	Astringent—Anti-Diarrheal—Treating Diabetes and Stomachache
Rutaceae	*Citrus bigaradia (C. aurantium)*	Evergreen Tree—Aromatic and Thick and Juicy Petal	Leaf—Flower—Fruit	Stachydrine—Volatile Oils—Hesperidin—Umbelliferone—Bergapten—Vitamins A, B$_3$, and C	Edible—Digestive—Anti-Paroxysm, Anti-Fungal, Anti-Bacterial, Anti-Spasmodic, Anti-Emetic, Anti-Tussive—Diaphoretic—Digestive—Expectorant
Rutaceae	*Citrus limetta*	Spiny, Small Tree—Evergreen—Brownish—Grey Bark—White Flower—Green-Yellow Fruit	Fruit	Vitamin C—Flavonoids—Hesperidin—Polyphenols	Edible—Treating Influenza and Colds—Bactericidal

(Continued)

TABLE 2.1 (CONTINUED)

Spermatophytes: Angiosperms, Dicotyledons, Dialypetalaes

Family Name	Species Name	Major Features of Botany	Usable Parts	Some Important Substances	Edible, Some Industrial and Medicinal Benefits
Rutaceae	*Citrus limonum* (*C. limon*)	Small Tree—Evergreen—Strong Root in White—Spiny Twig	Fruit	Vitamin C—Camphene—Limonene—Geraniol—Citral—Citric Acid—Flavonoids—Volatile Oils—Bergapten	Edible—Increasing White Globules—Anti-Rheumatism and Anti-Asthma—Tonic—Treating Scurvy
Rutaceae	*Citrus nobilis* (*C. reticulata*)	Tree—Spiny Branch—Evergreen	Fruit	Vitamin C—Limonoids—Nobiletin—Bergapten—Carotene—Flavonoids—Volatile Oils	Edible—Painkiller—Antiseptic, Anti-Emetic, Anti-Inflammatory, Anti-Scorbutic, Anti-Tussive—Aphrodisiac—Astringent—Laxative—Tonic
Rutaceae	*Citrus paradisi*	Tree—Evergreen—Big and Sour Fruit	Fruit—Bark—Flower	Kaempferol—Pectin—Lycopene—Vitamin C	Edible—Diuretic—Treating Blood Pressure—Anti-Depressant, Anti-Bacterial, Anti-Atherogenic, Anti-Inflammatory, Anti-HIV
Rutaceae	*Citrus sinensis*	Small Tree—White Flower with Weak Smell	Fruit	Vitamin C—Flavonoids—Volatile Oils—Bergapten	Edible—Antiseptic—Mucus Producing—Disinfectant—Dialysis
Saxifragaceae	*Ribes rubrum*	Shrub—White Flower—Red Fruit	Fruit—Leaf	Cyanidin—Organic Acids—Glucosides	Edible—Appetizer—Diuretic—Treating Rheumatism—Anti-Scorbutic—Aperient—Depurative—Digestive—Laxative—Refrigerant—Sialagogue
Tamaricaceae	*Tamarix gallica*	Bushy Tree—Narrow and Dark Green Leaf, Dark Red Branch	Manna—Gall—Leaf—Branchlet	Gallic Acid	Astringent—Detergent—Expectorant—Laxative—Diuretic—Appetizer—Treating Wounds, Diarrhea, and Dysentery

(Continued)

TABLE 2.1 (CONTINUED)

Spermatophytes: Angiosperms, Dicotyledons, Dialypetalaes

Family Name	Species Name	Major Features of Botany	Usable Parts	Some Important Substances	Edible, Some Industrial and Medicinal Benefits
Tropaeolaceae	*Tropaeolum majus*	Herbal—Big Flower in Yellow	Whole Plant—Leaf	Glucotropaeolin—Glycosides	Dialysis—Disinfectant—Aperient—Expectorant—Laxative—Stimulant Depurative—Diuretic—Emmenagogue—Antibiotic, Anti-Bacterial, Anti-Fungal, Antiseptic, and Anti-Scurvy—Treating Wounds
Umbelliferae (Apiaceae)	*Anethum graveolens*	Herbal—White Root—Small and Yellow Flower	Leaf—Fruit	Limonene—Carvone—Phellandrene	Edible—Anti-Nausea and Stomachache
Umbelliferae (Apiaceae)	*Apium graveolens*	Herbal—Small Flower in White-Green	Whole Plant	Apiin—Apiose	Edible—Diuretic—Treating Bloating
Umbelliferae (Apiaceae)	*Carum carvi*	Herbal—Small Aromatic Flower in White—Fleshy Root	Seed—Fruit	Wax—Mucilage—Carvone—Dihydrocarveol	Diuretic—Treating Bloating and Paroxysm
Umbelliferae (Apiaceae)	*Carum bulbocastanum (C. persicum, Bunium bulbocastanum, B. persicum)*	Herbal—White Flower	Seed—Fruit	Carvone—Flavonoids—Quercetin—Cuminaldehyde—Cymene—Limonene—Pinene—Terpinene	Astringent—Treating Bloating
Umbelliferae (Apiaceae)	*Coriandrum sativum*	Herbal—Full of Leaves in Green—Big and White Flower	Seed—Stem—Fruit—Leaf	Phellandrene—Coriandrol	Edible—Diuretic—Treating Bloating and Paroxysm—Expectorant
Umbelliferae (Apiaceae)	*Cuminum cyminum*	Herbal—White Root and Tall	Seed—Fruit	Phellandrene—Cuminol—Cumin Aldehyde—Cymene—Carvone	Stomach Tonic—Diuretic—Treating Bloating, Paroxysm, and Epilepsy

(Continued)

TABLE 2.1 (CONTINUED)

Spermatophytes: Angiosperms, Dicotyledons, Dialypetalaes

Family Name	Species Name	Major Features of Botany	Usable Parts	Some Important Substances	Edible, Some Industrial and Medicinal Benefits
Umbelliferae (Apiaceae)	*Daucus carota*	Herbal—Small Flower in Yellow	Fruit—Root	Limonin—Pinene—Falcarinol—Vitamin A—Carotene—Falcarindol—Asarone—Choline—Ethanol—Formic Acid—Butyric Acid—Limonene—Malic Acid—Maltose—Oxalic Acid—Palmitic Acid—Pyrrolidine—Quinic Acid	Edible—Diuretic—Skin Hygiene—Anthelmintic—Carminative—Deobstruent—Galactogogue—Ophthalmic—Stimulant
Umbelliferae (Apiaceae)	*Dorema ammoniacum*	Herbal—Covered by Trichome—Very Small and White Flower	Resin	Ferrolin—Saponins—Flavonoids—Alkaloids—Tannins—Resin	Industrial (Porcelain Making)—Edible—Tonic—Emmenagogue
Umbelliferae (Apiaceae)	*Foeniculum vulgare*	Herbal—Aromatic—Yellow Flower	Root—Leaf—Fruit—Seed	Anethole—Estragole—Phellandrene	Diuretic—Sedative—Appetizer—Emmenagogue—Analgesic—Carminative—Expectorant—Galactogogue—Hallucinogenic—Laxative—Stimulant—Stomachic—Anti-Inflammatory and Anti-Spasmodic
Umbelliferae (Apiaceae)	*Ferula gummosa*	Herbal—Thick Stem—Yellow Flower—Leaf with Trichome	Plant Juice—Root—Whole Plant	Cadinene—Cadinol—Umbelliferone—Galbanum	Industrial (Adhesive Making)—Treating Paroxysm and Stomachache—Anti- Spasmodic—Carminative—Expectorant—Stimulant
Umbelliferae (Apiaceae)	*Heracleum persicum*	Herbal—Aromatic—White Flower	Root—Stem—Flower	Methyl Butyrate—Ethyl Butyrate—Acetate—Furanocoumarin	Edible—Treating Bloating, Convulsions, Inflammation, and Fungal Diseases

(Continued)

TABLE 2.1 (CONTINUED)

Spermatophytes: Angiosperms, Dicotyledons, Dialypetalaes

Family Name	Species Name	Major Features of Botany	Usable Parts	Some Important Substances	Edible, Some Industrial and Medicinal Benefits
Umbelliferae (Apiaceae)	*Petroselinum crispum*	Herbal—Small and Green Flower	Whole Plant	Apioside—Apiiose—Apiin	Edible—Diuretic—Aperient—Carminative—Digestive—Emmenagogue—Expectorant—Galactofuge—Stomachic—Tonic—Treating Shortness of Breath and Hepatitis—Anti-Dandruff and Anti-Spasmodic
Umbelliferae (Apiaceae)	*Pimpinella anisum*	Herbal—Small Flower in White	Seed	Anisic Acid—Estragole—Anethole	Treating Bloating—Carminative—Digestive—Expectorant—Pectoral—Stimulant—Stomachic—Tonic—Antiseptic and Anti-Spasmodic
Violaceae	*Viola odorata*	Herbal—Heart-Like Leaf—Single Violet Flower—Frout with Trichome	Flower—Leaf—Root—Seed	Mucilage—Salicylic Acid—Anthocyanin	Softener—Demulcent—Emollient—Sweaty—Mucus Producing—Treating Bronchitis, Respiratory Catarrh, Coughs, Asthma, and Cancer of the Breast
Vitaceae (Ampelidaceae)	*Vitis vinifera*	Climber Bushy Tree—Knotted Stem	Fruit—Leaf—Seed—Sap—Tendril	Succinic Acid—Glycolic Acid	Edible—Tonic—Astringent—Diuretic—Constructive—Cooling—Demulcent—Expectorant—Laxative—Stomachic—Treating Fever Blister, Varicose Veins, Hemorrhoids, Diarrhea, and Capillary Fragility—Anti-Lithic and Anti-Inflammatory

(Continued)

TABLE 2.1 (CONTINUED)

Spermatophytes: Angiosperms, Dicotyledons, Dialypetalaes

Family Name	Species Name	Major Features of Botany	Usable Parts	Some Important Substances	Edible, Some Industrial and Medicinal Benefits
Zygophyllaceae	*Peganum harmala*	Herbal—Full of Green Leaves—Big White Flower	Seed—Fruit—Leaf—Root	Harmin—Harmaline—Harmalol	Hypnotic—Digestive—Diuretic—Hallucinogenic—Narcotic—Stimulant—Sweaty—Galactogogue—Ophthalmic—Soporific—Vermifuge—Anti-Worm
Zygophyllaceae	*Tribulus terrestris*	Herbal—Stem with Hair	Fruit—Flower—Stem—Seed	Resin—Polyphenols—Arabinose	Tonic—Appetizer—Bladder and Kidney Stone Exertion—Abortifacient—Alterative—Anthelmintic—Aphrodisiac—Astringent—Carminative—Demulcent—Diuretic—Emmenagogue—Galactogogue—Pectoral—Treating Congestion, Headache, Liver, Ophthalmia, and Stomatitis

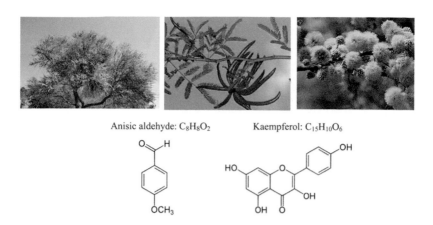

Anisic aldehyde: $C_8H_8O_2$ Kaempferol: $C_{15}H_{10}O_6$

FIGURE 2.1 *Acacia farnesiana* (shrub, green seeds, and flowers) and its main components: Anisic aldehyde, kaempferol.

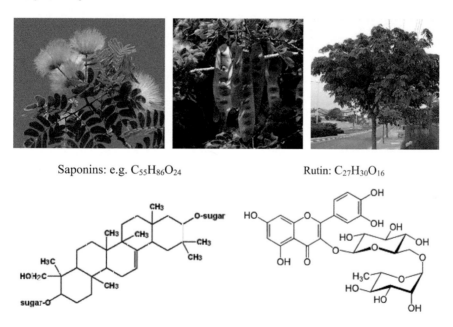

Saponins: e.g. $C_{55}H_{86}O_{24}$ Rutin: $C_{27}H_{30}O_{16}$

FIGURE 2.2 *Albizia lebbeck* (flowers, pods, and tree) and its main components: Saponins, rutin.

Alhaji camelorum (Figure 2.3) is a deciduous bushy tree that grows up to a size of 2 m. The flowers are hermaphrodite. It can fix nitrogen. It has a preference for light (sandy), medium (loamy), and well-drained soil, and can grow in acidic, neutral, basic (alkaline), and saline soils. It cannot grow in the shade. It prefers dry or moist soil.

The whole plant is diaphoretic, diuretic, expectorant, and laxative. An oil from the leaves is used in the treatment of rheumatism. The flowers are used in the treatment of piles (Zargari, 2014; Mozaffarian, 2011; pfaf.org; Laghari *et al.*, 2010).

Althaea officinalis (Figure 2.4) is a perennial that grows to a size of 1.2 m by 0.8 m. The flowers are hermaphrodite and are pollinated by bees. The plant is self-fertile.

Melezitose: $C_{18}H_{32}O_{16}$

FIGURE 2.3   *Alhagi camelorum* (flowers, shrub, and leaves) and its main component: Melezitose.

FIGURE 2.4   *Althaea officinalis* (leaf, flowers, and whole plant).

It has a preference for light (sandy), medium (loamy), and heavy (clay) soils, and can grow in acidic, neutral, basic (alkaline), and saline soils. It cannot grow in the shade. It prefers dry or moist soil.

"Marshmallow" is a very useful household medicinal herb. Its soothing demulcent properties make it very effective in treating inflammation and irritation of the mucous membranes such as the alimentary canal, and the urinary and respiratory organs. The roots counter excess stomach acid, peptic ulceration, and gastritis. It is also applied externally to bruises, sprains, aching muscles, insect bites, skin inflammations, splinters, etc. The whole plant, but especially the root, is anti-tussive, demulcent, diuretic, highly emollient, slightly laxative, and odontalgic. An infusion of the leaves is used to treat cystitis and frequent urination. The leaves are harvested in August when the plant is just coming into flower and can be dried for later use. The root can be used in an ointment for treating boils and abscesses. The root is best harvested in the autumn, preferably from two-year-old plants, and is dried for later use. The German Commission E Monographs: Therapeutic guide to herbal medicines

(Blumenthal *et al.*, 1998), approve marshmallow for the treatment of irritation of mouth and throat and associated dry cough, bronchitis (root and leaf), and mild stomach lining inflammation (root). No documented adverse effects are reported but anecdotal reports suggest an allergic reaction and lowering of blood sugar (Zargari, 2014; Mozaffarian, 2011; pfaf.org; Al Snafi, 2013).

Amygdalus communis (Figure 2.5) is a deciduous tree that grows up to a size of 6 m. The flowers are hermaphrodite and pollinated by insects. The plant is self-fertile. It has a preference for light (sandy), medium (loamy), heavy (clay), and well-drained soil, and prefers acidic, neutral, and basic (alkaline) soils. It cannot grow in the shade. It prefers moist soil.

As well as being a tasty addition to the diet, "almonds" are also beneficial to the overall health of the body, being used especially in the treatment of kidney stones, gallstones, and constipation. Externally, the oil is applied to dry skin and is also often used as a carrier oil in aromatherapy. The seed is demulcent, emollient, laxative, nutritive, and pectoral. When used medicinally, the fixed oil from the seed is normally employed. The seed contains "laetrile," a substance that has also been called amygdalin or its misnomer, vitamin B_{17} ($C_{20}H_{27}NO_{11}$). This is claimed to have a positive effect in the treatment of cancer, but at present there does not seem to be much evidence to support this. The pure substance is almost harmless, but on hydrolysis it yields hydrocyanic acid (HCN), a very rapidly acting poison—it should thus be treated with caution. In small amounts this exceedingly poisonous compound stimulates respiration, improves digestion, and gives a sense of well-being. The leaves are used in the treatment of diabetes. The plant contains the anti-tumor compound taxifolin ($C_{15}H_{12}O_7$). Although no specific mention has been made of this species, it belongs to a genus where most, if not all members of the genus, produce hydrocyanic acid, a poison that gives almonds their characteristic flavor. This toxin is found mainly in the leaves and seed and is readily detected by its bitter taste. It is usually present in too small a quantity to do any harm, but any very bitter seed or fruit should not be eaten. In small quantities, hydrocyanic acid (HCN) has been shown to stimulate respiration and improve digestion; as mentioned, it is also claimed to be of benefit in the treatment of cancer. In excess, however, it can cause respiratory failure and even death (Zargari, 2014; Mozaffarian, 2011; pfaf.org; Moosavi *et al.*, 2014).

Anethum graveolens (Figure 2.6) is an annual that grows to a size of 0.8 m by 0.2 m at a medium rate. The flowers are hermaphrodite and pollinated by bees. The plant is self-fertile. It is noted for attracting wildlife. It has a preference for light (sandy), medium (loamy), and well-drained soil, and prefers acidic, neutral, and basic (alkaline) soils. It cannot grow in the shade. It prefers moist soil.

"Dill" has a very long history of herbal use going back more than 2,000 years. The seeds are a common and very effective household remedy for a wide range of digestive problems. An infusion is especially efficacious in treating gripe in babies and flatulence in young children. The seed is aromatic, carminative, mildly diuretic, galactogogue, stimulant, and stomachic. It is also used in the form of an extracted essential oil. Used either in an infusion, or by eating the seed whole, the essential oil in the seed relieves intestinal spasms and griping, helping to settle colic. Chewing the seed improves bad breath. Dill is also a useful addition to cough, cold, and flu remedies; it can be used with anti-spasmodics such as *Viburnum opulus* to relieve period pains. Dill also helps to increase the flow of milk in nursing mothers and is then taken by the baby in the milk to help prevent colic. Dill is said to contain the alleged

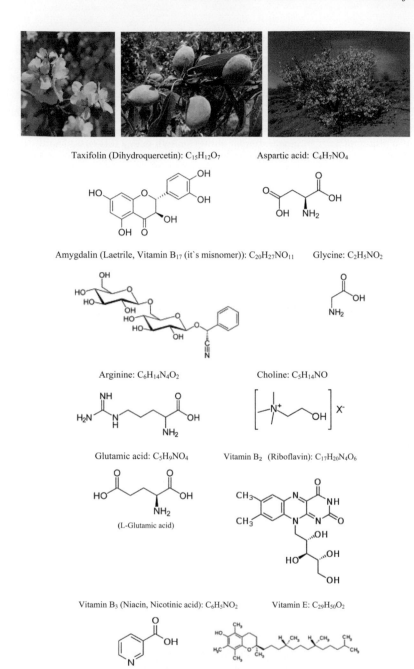

FIGURE 2.5 *Amygdalus communis* (blossoms, fruits, and tree) and its main components: Taxifolin, aspartic acid, amygdalin, glycine, arginine, choline, glutamic acid, vitamins B_2, B_3, E.

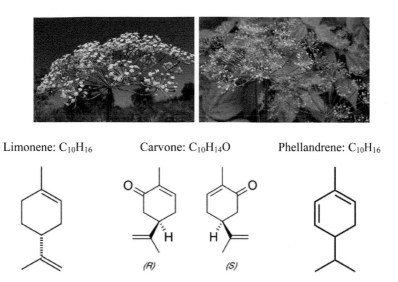

Limonene: C$_{10}$H$_{16}$ Carvone: C$_{10}$H$_{14}$O   Phellandrene: C$_{10}$H$_{16}$

FIGURE 2.6 *Anethum graveolens* (flowers and whole plant) and its main components: Limonene, carvone, phellandrene.

"psychotroph" myristicine. There are also reports that dill can cause photosensitivity and/or dermatitis in some people. Dill oil should be avoided during pregnancy (Zargari, 2014; Mozaffarian, 2011; pfaf.org; Kaur & Arora, 2009).

Apium graveolens (Figure 2.7) is a biennial that grows to a size of 0.6 m by 0.3 m. The flowers are hermaphrodite and pollinated by flies. The plant is self-fertile. It has a preference for light (sandy), medium (loamy), and heavy (clay) soils, and can grow in acidic, neutral, basic (alkaline), and saline soils, and in semi-shade (light woodland). It prefers moist soil.

"Wild celery" has a long history of medicinal and food use. It is an aromatic bitter tonic herb that reduces blood pressure, relieves indigestion, stimulates the uterus, and is anti-inflammatory. The ripe seeds, herb, and root are aperient, carminative, diuretic, emmenagogue, galactogogue, nervine, stimulant, and tonic. Wild celery is said to be useful in cases of hysteria, promoting restfulness and sleep and diffusing a mild sustaining influence through the system. The herb should not be prescribed for pregnant women. Seeds purchased for cultivation purposes are often treated with a fungicide; these should not be used for medicinal purposes. The root is harvested in the autumn and can be used fresh or dried. The whole plant is harvested when fruiting and is usually liquidized to extract the juice. The seeds are harvested as they ripen and are dried for later use. An essential oil obtained from the plant has a calming effect on the central nervous system. Some of its constituents have anti-spasmodic, sedative, and anti-convulsant actions. It has been shown to be of value in treating high blood pressure. A homeopathic remedy is made from the herb, which is used in treating rheumatism and kidney complaints. If the plant is infected with the fungus *Sclerotinia sclerotiorum*, skin contact with the sap can cause dermatitis in people with sensitive skin. This is more likely to happen to Caucasians. Allergic responses include anaphylaxis in sensitive individuals. There is cross-allergenicity

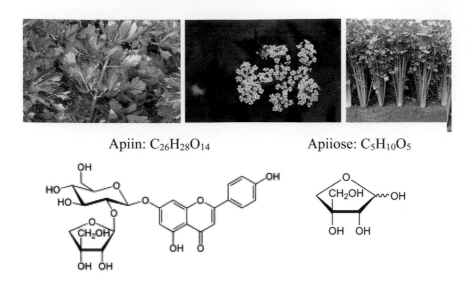

Apiin: $C_{26}H_{28}O_{14}$ Apiiose: $C_5H_{10}O_5$

FIGURE 2.7 *Apium graveolens* (leaves, flowers, and whole plant) and its main components: Apiin, apiiose.

between celery, cucumber, carrot, watermelon, and possibly apples. Avoid during pregnancy as emmenagogue, abortifacient, and uterine stimulant activity has been reported (Zargari, 2014; Mozaffarian, 2011; pfaf.org; Kooti & Daraei, 2017).

Aquilegia vulgaris (Figure 2.8) is a perennial that grows to a size of 1 m by 0.5 m at a medium rate. The flowers are hermaphrodite and pollinated by bees. It has a preference for light (sandy), medium (loamy), and well-drained soil. It can grow in acidic, neutral, and basic (alkaline) soils and in semi-shade (light woodland) or no shade. It prefers moist soil.

"Common columbine" was formerly employed in herbal medicine mainly for its anti-scorbutic effect, but it has fallen out of favor and is little used nowadays. The leaves, root, and seed are astringent, depurative, diaphoretic, diuretic, and parasiticide. Because of its toxic properties, this plant should not be taken internally without expert advice, though the root is sometimes used externally in poultices to treat ulcers

FIGURE 2.8 *Aquilegia vulgaris* (whole plant and flowers).

and the more common skin diseases. Columbine has produced very unsatisfactory results and is not normally used medicinally. A homeopathic remedy is made from the plant, which is used in treatment related to the nervous system. The plant is poisonous though the toxins are destroyed by heat or by drying. Although this plant contains alkaloids, no cases of poisoning to humans or other mammals have been recorded (Zargari, 2014; Mozaffarian, 2011; pfaf.org; Bylka *et al.*, 2004).

Arachis hypogaea (Figure 2.9) is an annual that grows up to a size of 0.3 m. The flowers are hermaphrodite and pollinated by insects. The plant is self-fertile. It can fix nitrogen. It has a preference for light (sandy), medium (loamy), heavy (clay), and well-drained soil. It can grow in acidic, neutral, and basic (alkaline) soils; it can also grow in very acidic and very alkaline soils. It cannot grow in the shade. It prefers moist soil.

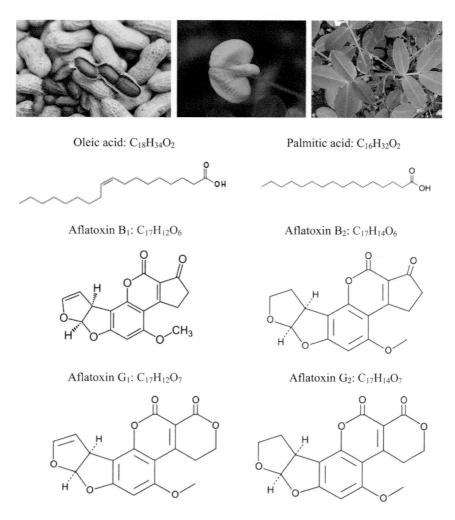

Oleic acid: $C_{18}H_{34}O_2$

Palmitic acid: $C_{16}H_{32}O_2$

Aflatoxin B_1: $C_{17}H_{12}O_6$

Aflatoxin B_2: $C_{17}H_{14}O_6$

Aflatoxin G_1: $C_{17}H_{12}O_7$

Aflatoxin G_2: $C_{17}H_{14}O_7$

FIGURE 2.9 *Arachis hypogaea* (seeds, flower, and leaves) and its main components: Oleic acid, palmitic acid, aflatoxins B_1, B_2, G_1, G_2.

The oil from the seed of the "peanut" is aperient, demulcent, emollient, and pectoral. The seed is used mainly as a nutritive food. The seeds have been used in folk medicine as an anti-inflammatory, aphrodisiac, and decoagulant. Peanuts play a small role in various folk pharmacopoeias. In China the nuts are considered demulcent, pectoral, and peptic; the oil is considered aperient and emollient, taken internally in milk for treating gonorrhea, and externally for treating rheumatism. In Zimbabwe the peanut is used in folk remedies for plantar warts. Hemostatic and vasoconstrictor activity are reported. The alcoholic extract is said to affect isolated smooth muscles and frog hearts like acetylcholine. The alcoholic lipoid fraction of the seed is said to prevent hemophilia tendencies and is said to be used in the treatment of some blood disorders (mucorrhagia and arthritic hemorrhages) in hemophiliacs. Of greatest concern is possible contamination of damaged or spoiled seeds with the teratogenic, carcinogenic aflatoxins. Two principal toxins, aflatoxins B_1 ($C_{17}H_{12}O_6$) and G_1 ($C_{17}H_{12}O_7$), and their less toxic dihydro derivatives, aflatoxins B2 ($C_{17}H_{14}O_6$) and G2 ($C_{17}H_{14}O_7$), are formed by the aflatoxin producing molds (*Aspergillus flavus et al.*). Prevention of mold growth is the mainstay, there being no satisfactory way to remove the toxins from feed and foods (however, peanut oils are free of aflatoxins because of alkaline processing). Avoid if there is any suggestion of an allergy (Zargari, 2014; Mozaffarian, 2011; pfaf.org; Al Snafi, 2014).

Armeniaca vulgaris (Prunus armeniaca) (Figure 2.10) is a deciduous tree that grows to a size of 9 m by 6 m at a medium rate. The flowers are hermaphrodite and pollinated by insects. The plant is self-fertile. It has a preference for light (sandy), medium (loamy), and well-drained soil. It can grow in acidic, neutral, and basic (alkaline) soils and semi-shade (light woodland) or no shade. It prefers moist soil.

"Apricot" fruits contain citric acid ($C_6H_8O_7$) and tartaric acid ($C_4H_6O_6$), carotenoids and flavonoids. They are nutritious, cleansing, and mildly laxative. They are a valuable addition to the diet, working gently to improve overall health. The salted fruit is anti-inflammatory and antiseptic. It is used medicinally in Vietnam in the treatment of respiratory and digestive diseases; it is also anti-pyretic, antiseptic, emetic, and ophthalmic. The flowers are tonic, promoting fecundity in women. The bark is astringent. The inner bark and/or the root are used for treating poisoning caused by eating bitter almond and apricot seeds (which contain HCN). Another report says that a decoction of the outer bark is used to neutralize the effects of HCN. The decoction is also used to soothe inflamed and irritated skin conditions. The seed is analgesic, anthelmintic, anti-asthmatic, anti-spasmodic, anti-tussive, demulcent, emollient, expectorant, pectoral, sedative, and vulnerary. It is used in the treatment of asthma, coughs, acute or chronic bronchitis, and constipation. The seed contains "laetrile," a substance that has also been called vitamin B_{17}. This has been claimed to have a positive effect in the treatment of cancer, but at present there does not seem to be much evidence to support this. The pure substance is almost harmless, but on hydrolysis it yields HCN, a very rapidly acting poison—it should thus be treated with caution. In small amounts this exceedingly poisonous compound stimulates respiration, improves digestion, and gives a sense of well-being. This species produces HCN, a poison that gives apricots their characteristic flavor. This toxin is found mainly in the leaves and seed and is readily detected by its bitter taste. Usually present in too small a quantity to do any harm, any very bitter seed or fruit should not be eaten. In small quantities, HCN has been shown to stimulate respiration and improve digestion; as mentioned, it is also claimed to be of benefit in the treatment of cancer. In excess, however, it can cause

Pangamic acid (Pangamate): $C_{10}H_{19}NO_8$

Citric acid: $C_6H_8O_7$ Tartaric acid: $C_4H_6O_6$

FIGURE 2.10 *Armeniaca vulgaris* (flowers, fruits, and tree) and its main components: Pangamic acid, citric acid, tartaric acid, and amygdalin (not shown).

respiratory failure and even death. Oral doses of 50 g of HCN can be fatal (= 30 g of kernels or 50–60 kernels at 2 mg HCN/g kernel) (Zargari, 2014; Mozaffarian, 2011; pfaf.org; Mozaffarian, 1996; Orhan & Karta, 2011).

Astragalus sp. (Figure 2.11) is a large genus of over 3000 species, belonging to the legume family and the faboideae subfamily. It is the largest genus of plants in terms of described species. The genus is native to temperate regions of the Northern Hemisphere. Iran has more than 900 species of the Astragalus genus. Due to the fact that it has many species, the Astragalus genus has different characteristics in terms of botany. In general terms, it can be stated that the plants are annual or perennial, herbaceous or shrub or bushy tree, single leaf or multi-leaves with either even or odd pair, the flowers are either spikes, cylindrical, or stacked casings with round or cylindrical grains, and it also has highly variable fruits.

The natural gum tragacanth is made from several species of "milkvetch" occurring in the Middle East, including *A. adscendens*, *A. gummifer*, *A. brachycalyx*, and *A. tragacanthus*. Also *A. propinquus* (syn. *A. membranaceus*) has a history of use as a herbal medicine used in systems of traditional Chinese and Persian medicine. Biotechnology firms are working on deriving a telomerase activator from Astragalus. The chemical constituent cycloastragenol (also called TAT2) ($C_{30}H_{50}O_5$) is being studied to help combat HIV, as well as infections associated with chronic diseases or

Cycloastragenol: C$_{30}$H$_{50}$O$_5$

FIGURE 2.11 *Astragalus sp.* (whole plant, flowers, and fruits) and its main component: Cycloastragenol.

aging. Astragalus may interact with medications that suppress the immune system, such as cyclophosphamide. It may also affect blood sugar levels and blood pressure. Some Astragalus species can be toxic (Zargari, 2014; Mozaffarian, 2011; pfaf.org; Niknam & Salehi Lisar, 2004; Owfi & Barani, 2017).

Berberis vulgaris (Figure 2.12) is a deciduous bushy tree that grows to a size of 3 m by 2 m at a medium rate. The flowers are hermaphrodite and pollinated by insects. The plant is self-fertile. It is noted for attracting wildlife. It has a preference for light (sandy), and medium (loamy), and can grow in heavy clay and nutritionally poor soils. It can grow in acidic, neutral, and basic (alkaline) soils, and semi-shade (light woodland) or no shade. It prefers dry or moist soil.

"Barberry" has long been used as a herbal remedy for the treatment of a variety of complaints. All parts of the plant can be used, though the yellow root bark is the most concentrated source of active ingredients. The plant is mainly used nowadays as a tonic to the gallbladder to improve the flow of bile and for ameliorate conditions such as gallbladder pain, gallstones, and jaundice. The bark and root bark are antiseptic, astringent, cholagogue, hepatic, purgative, refrigerant, stomachic, and tonic. The bark is harvested in the summer and can be dried for storing. It is especially useful in cases of jaundice, general debility, and biliousness, but should be used with caution. The flowers and the stem bark are anti-rheumatic. The roots are astringent and antiseptic, which are pulverized in a little water and used to treat mouth ulcers. A tea from the roots and stems has been used to treat stomach ulcers. The root bark has also been used as a purgative and treatment for diarrhea and is diaphoretic. A tincture of the root bark has been used in the treatment of rheumatism, sciatica, etc. The root bark is a rich source of the alkaloid berberine—about 6%. Berberine, universally present in rhizomes of the berberis species, has marked anti-bacterial effects. Since it is not appreciably absorbed by the body, it is used orally in the treatment of various enteric infections, especially bacterial dysentery. It should not be used with

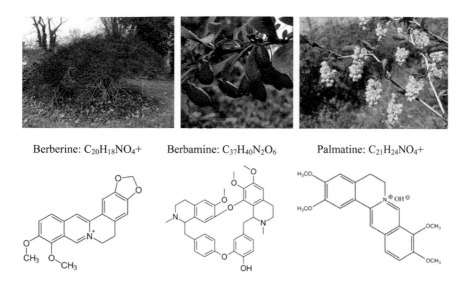

Berberine: $C_{20}H_{18}NO_4+$ Berbamine: $C_{37}H_{40}N_2O_6$ Palmatine: $C_{21}H_{24}NO_4+$

FIGURE 2.12 *Berberis vulgaris* (shrub, fruits, and flowers) and its main components: Berberine, berbamine, palmatine.

the glycyrrhiza species (licorice) because this nullifies the effects of the berberine ($C_{20}H_{18}NO_4+$). Berberine ($C_{20}H_{18}NO_4+$) has also shown anti-tumor activity and is also effective in the treatment of hypersensitive eyes, inflamed lids, and conjunctivitis. A tea made from the fruits is antiseptic, appetizer, astringent, diuretic, expectorant, and laxative. It is also used as a febrifuge. The fruit, or freshly pressed juice, is used in the treatment of liver and gall bladder problems, kidney stones, menstrual pains, etc. The leaves are astringent and anti-scorbutic. A tea made from the leaves is used in the treatment of coughs. The plant (probably the inner bark) is used by homeopaths as a valuable remedy for kidney and liver insufficiency. Other uses include malaria, and opium and morphine withdrawal. The bark can be used in doses of 4 mg or more to treat stupor, nosebleeds, vomiting, diarrhea, and kidney irritation. It is contraindicated during pregnancy as there is an abortion risk (Zargari, 2014; Mozaffarian, 2011; pfaf.org; Rahimi Madiseh *et al.*, 2017).

Brassica napus (Figure 2.13) is an annual/biennial that grows up to a size of 1.2 m. The flowers are hermaphrodite and pollinated by bees. The plant is self-fertile. It has a preference for light (sandy), medium (loamy), and well-drained soil. It can grow in heavy clay soil, acidic, neutral, and basic (alkaline) soils, as well as in very acidic and very alkaline soils. In addition, it can grow in semi-shade (light woodland) or no shade. It prefers moist soil.

The root of "rapeseed" is emollient and diuretic. The juice of the roots is used in the treatment of chronic coughs and bronchial catarrh. The seed, powdered and with salt, is said to be a folk remedy for cancer. Rape oil is used in massages and oil baths as it is believed to strengthen the skin and keep it cool and healthy. With camphor it is applied as a remedy for rheumatism and stiff joints. It is dropped into the ear to relieve earaches. The oil contained in the seed of some varieties of this species can be rich in erucic acid which is toxic. However, modern cultivars have been selected which are almost free of erucic acid (Zargari, 2014; Mozaffarian, 2011; pfaf.org; Saeidnia & Gohari, 2012).

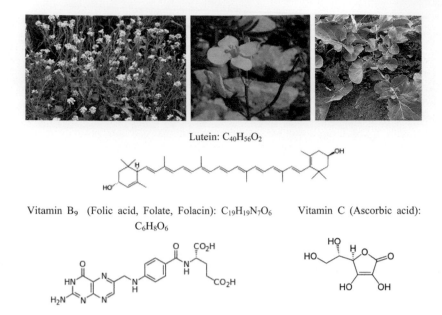

Lutein: $C_{40}H_{56}O_2$

Vitamin B_9 (Folic acid, Folate, Folacin): $C_{19}H_{19}N_7O_6$ Vitamin C (Ascorbic acid):
$C_6H_8O_6$

FIGURE 2.13 *Brassica napus* (whole plant, flower, and leaves) and its main components: Lutein, vitamins B_9, C.

Brassica nigra (Figure 2.14) is an annual that grows to a size of 1.2 m by 0.6 m. The species is hermaphrodite and pollinated by bees and flies. The plant is self-fertile. It has a preference for light (sandy) medium (loamy), and well-drained soil. It can grow in acidic, neutral, and basic (alkaline) soils as well as very acidic soils, also in semi-shade (light woodland) or no shade. It prefers moist soil. The plant can tolerate maritime exposure.

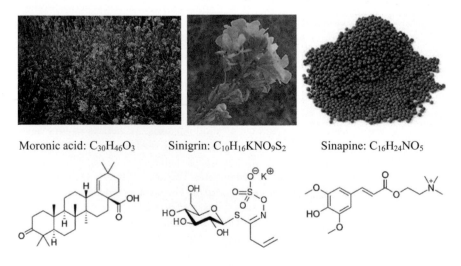

Moronic acid: $C_{30}H_{46}O_3$ Sinigrin: $C_{10}H_{16}KNO_9S_2$ Sinapine: $C_{16}H_{24}NO_5$

FIGURE 2.14 *Brassica nigra* (whole plant, flower, and seeds) and its main components: Moronic acid, sinigrin, sinapine.

The "black mustard" seed is often used in herbal medicine, especially as a rubefacient poultice. The seed is ground and made into a paste and then applied to the skin in the treatment of rheumatism, as a means of reducing congestion in the internal organs. Applied externally, mustard relieves congestion by drawing the blood to the surface as in head afflictions, neuralgia, and spasms. Hot water poured on bruised seeds makes a stimulant foot bath, and is good for colds and headaches. Old herbal medicines suggest mustard for treating alopecia, epilepsy, snake bites, and toothache. Care must be taken not to overdo it, since poultices can sometimes cause quite severe irritation to the skin. The seed is also used internally, where it is appetizer, digestive, diuretic, emetic, and tonic. Swallowed whole when mixed with molasses, it acts as a laxative. A decoction of the seeds is used in the treatment of indurations of the liver and spleen. It is also used to treat carcinoma, throat tumors, and imposthumes. A liquid prepared from the seed, when gargled, is said to help tumors of the sinax. The seed is eaten as a tonic and appetite stimulant. Hot water poured onto bruised mustard seeds makes a stimulating foot bath and can also be used as an inhaler where it acts to throw off a cold or dispel a headache. Mustard oil is said to stimulate hair growth. Mustard is also recommended as an aperient ingredient of tea, useful for hiccups. Mustard flour is considered antiseptic (Zargari, 2014; Mozaffarian, 2011; pfaf.org; Krishnaveni & Saranya, 2016).

Brassica oleracea (Figure 2.15) is a biennial that grows up to a size of 0.8 m. The flowers are hermaphrodite and pollinated by bees. The plant is self-fertile. It has a preference for light (sandy), medium (loamy), and well-drained soil. It tolerates heavy clay soil. It can grow in acidic, neutral, and basic (alkaline) soils, as well as in semi-shade (light woodland) or no shade. It prefers moist soil. The plant can tolerate maritime exposure.

"Cabbage" is used for stomach pain, excess stomach acid, stomach and intestinal ulcers, and a stomach condition called Roemheld syndrome. Cabbage is also used to treat asthma and morning sickness. Cabbage is also used to prevent weak bones (osteoporosis), as well as cancer of the lung, stomach, colon, breast, and other types of cancer. Breast-feeding women sometimes apply cabbage leaves and cabbage leaf

Sulforaphane: $C_6H_{11}NOS_2$

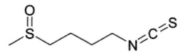

FIGURE 2.15 *Brassica oleracea* (fruit, flowers, and leaves) and its main components: Sulforaphane, and vitamin C (not shown).

extracts to their breasts to relieve swelling and pain. The herbal poultice for the treatment of external injuries is the best known of all cabbage-derived herbal medications. This form of the remedy is prepared using the leaves of the wild or cultivated cabbages. To aid in the detoxification of the liver, the leaves of the wild cabbage are often eaten raw or they may be consumed cooked as an aid to digestion. The detoxification effect of the cabbage is significant; at the same time, the vegetable is considered very helpful in the treatment of painful arthritis over the long term. The vitamin C ($C_6H_8O_6$) deficiency disease called scurvy can be beaten back and cured by eating raw cabbage rich in vitamin C ($C_6H_8O_6$). Packed cabbage leaves were often used as topical eczema remedies in many parts of Europe in earlier times. As a topical remedy, cabbage was also used for the treatment of different disorders affecting the legs— including varicose veins and in the alleviation of leg ulcers of all sorts. When used as internal as well as external remedies, cabbage is useful mainly due to its high sulfur content; the mineral destroys the ferments within the blood, and this is particularly effective as a treatment for skin disorders of all kinds. Individuals affected by persistent coldness in the feet can benefit by including cabbage in their daily diet, because sulfur is considered to be one of those elements that can induce an increase in the production of heat in the body. Cabbage might affect blood sugar levels in people with diabetes. Watch for signs of low blood sugar (hypoglycemia) and monitor your blood sugar carefully if you have diabetes and use cabbage (Zargari, 2014; Mozaffarian, 2011; pfaf.org; Maggioni *et al.*, 2018).

Carum bulbocastanum (*C. persicum*, *Bunium bulbocastanum*, *B. persicum*) (Figure 2.16) is a perennial that grows up to a size of 0.6 m. The species is hermaphrodite and pollinated by insects. The plant is self-fertile. It has a preference for light (sandy), medium (loamy), heavy (clay), and well-drained soil. It can grow in acidic, neutral, and basic (alkaline) soils, and semi-shade (light woodland) or no shade. It prefers moist soil.

The chemical composition of "black cumin" has not been worked out in detail. The principal constituents of the essential oil are cuminaldehyde ($C_{10}H_{12}O$) (45.4%) and o-cymene ($C_{10}H_{14}$) (35%). Carvone ($C_{10}H_{14}O$), limonene ($C_{10}H_{16}$), pinene ($C_{10}H_{16}$), and terpinene ($C_{10}H_{16}$) are the minor constituents. It is astringent and is useful for treating bloating (Zargari, 2014; Mozaffarian, 2011; pfaf.org; Mozaffarian, 1996; Kapoor *et al.*, 2010).

Carum carvi (Figure 2.17) is a biennial that grows to a size of 0.6 m by 0.3 m. It is not frost tender. The flowers are hermaphrodite and pollinated by bees. The plant is self-fertile. It has a preference for light (sandy), medium (loamy), heavy (clay), and well-drained soil. It can grow in acidic, neutral, and basic (alkaline) soils, and semi-shade (light woodland) or no shade. It prefers moist soil.

"Caraway" has a long history of use as a household remedy especially in the treatment of digestive complaints where its anti-spasmodic action soothes the digestive tract and its carminative action relieves bloating caused by wind and improves the appetite. It is often added to laxative medicines to prevent griping. The seed is antiseptic, anti-spasmodic, aromatic, carminative, digestive, emmenagogue, expectorant, galactogogue, and stimulant. It can be chewed raw for the almost immediate relief of indigestion and can also be made into infusions. The seed is also used in the treatment of bronchitis and is found as an ingredient of cough remedies, especially useful for children. The seed is also said to increase the production of breast milk in nursing mothers. The seed is harvested when fully ripe, then dried and stored in a cool, dry

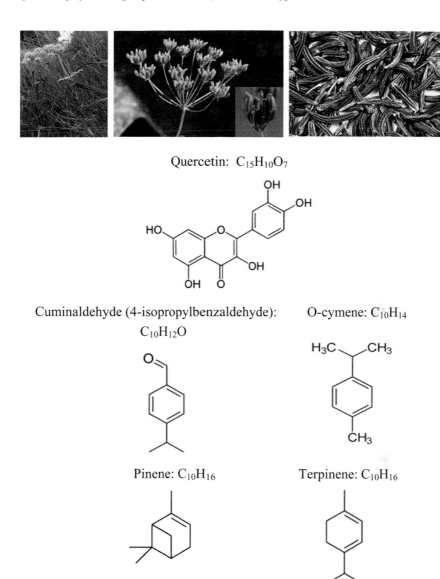

Quercetin: $C_{15}H_{10}O_7$

Cuminaldehyde (4-isopropylbenzaldehyde): $C_{10}H_{12}O$

O-cymene: $C_{10}H_{14}$

Pinene: $C_{10}H_{16}$

Terpinene: $C_{10}H_{16}$

(α-Terpinene)

FIGURE 2.16 *Carum bulbocastanum* (whole plant, flower, and seeds) and its main components: Quercetin, cuminaldehyde, o-cymene, pinene, terpinene.

place out of the sunlight. The essential oil can be extracted from the seed and has similar properties. A tea made from the seeds is a pleasant stomachic and carminative, and has been used to treat flatulent colic. The seed is used in Tibetan medicine where it is considered to have an acrid taste and a heating potency. It is used to treat failing vision and loss of appetite. Caraway is said to contain the alleged "psychotroph" myristicine. Excessive intake can lead to kidney and liver damage (Zargari, 2014; Mozaffarian, 2011; pfaf.org; Mohiyuddin Khan *et al.*, 2016).

Dihydrocarveol: $C_{10}H_{18}O$

FIGURE 2.17 *Carum carvi* (flowers, leaves, and whole plant) and its main components: Dihydrocarveol, and carvone (not shown).

Cassia fistula (Figure 2.18) is a deciduous tree that grows to a size of 18 m by 16 m at a slow rate. The flowers are pollinated by insects. It has a preference for light (sandy), medium (loamy), heavy (clay), and well-drained soil. It can grow in acidic, neutral, and saline soils. It cannot grow in the shade. It prefers moist soil and can tolerate drought. The plant can tolerate strong winds but not maritime exposure. Ripe pods and seeds are widely used in both traditional and conventional medicine as a laxative.

Anthraquinone: $C_{14}H_8O_2$

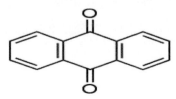

FIGURE 2.18 *Cassia fistula* (flowers, fruit, and tree) and its main component: Anthraquinone.

The root bark, leaves, and flowers of the "golden shower tree" have laxative properties, but to a lesser extent. In modern medicine, the fruit pulp is sometimes used as a mild laxative in pediatrics. The fruit pulp and leaves are rich in anthraquinone ($C_{14}H_8O_2$) derivatives (around 2%), and glycosides, which are responsible for the laxative properties. The fruit pulp is rich in pectins and mucilage. In-vitro and in-vivo tests have shown that the seed powder has amebicidal and cysticidal properties against *Entamoeba histolytica* and that it could cure intestinal amebiasis of humans. The aqueous fraction of the pods has produced a significant decrease in glycemia. Aqueous and methanolic bark extracts have shown significant antioxidant and anti-inflammatory activities. An alcohol extract of the leaves has shown anti-bacterial activity in-vivo against *Staphylococcus aureus* and *Pseudomonas aeruginosa*, plus accelerated wound healing. The pods are used as a remedy for malaria, blood poisoning, anthrax, diabetes, and dysentery (Zargari, 2014; Mozaffarian, 2011; pfaf.org; Pawar *et al.*, 2017).

Cerasus vulgaris (Prunus cerasus) (Figure 2.19) is a deciduous tree that grows up to a size of 6 m. The flowers are hermaphrodite and pollinated by bees. The plant is self-fertile. It has a preference for light (sandy), medium (loamy), heavy (clay), and well-drained soil. It can grow in acidic, neutral, and basic (alkaline) soils, including very acidic soils, also in semi-shade (light woodland) or no shade. It prefers moist soil. The plant can tolerate maritime exposure.

The bark of the "sour cherry" is astringent, bitter, and febrifuge. An infusion of the bark has been used in the treatment of fevers, coughs, and colds. The root bark has been used as a wash for cold sores and ulcers. The seed is nervine. Although no specific mention has been made of this species, all members of the genus contain amygdalin ($C_{20}H_{27}NO_{11}$) and prunasin ($C_{14}H_{17}NO_6$) substances which break down in

Prunasin: $C_{14}H_{17}NO_6$

FIGURE 2.19 *Cerasus vulgaris* (fruits and blossoms) and its main component: Prunasin.

water to form hydrocyanic acid (HCN). In small quantities, HCN has been shown to stimulate respiration, improve digestion, and give a sense of well-being. In addition, although no specific mention has been made of this species, it belongs to a genus where most, if not all members of the genus, produce hydrocyanic acid (HCN), a poison that gives almonds their characteristic flavor. This toxin is found mainly in the leaves and seed and is readily detected by its bitter taste. It is usually present in too small a quantity to do any harm, but any very bitter seed or fruit should not be eaten. In small quantities, hydrocyanic acid (HCN) has been shown to stimulate respiration and improve digestion; it is also claimed to be of benefit in the treatment of cancer. In excess, however, it can cause respiratory failure and even death (Zargari, 2014; Mozaffarian, 2011; pfaf.org; Mozaffarian, 1996; Piccolella *et al.*, 2008).

Cerasus avium (Prunus avium) (Figure 2.20) is a deciduous tree that grows to a size of 18 m by 7 m at a fast rate. The flowers are hermaphrodite and pollinated by bees. The plant is not self-fertile. It is noted for attracting wildlife. It has a preference for light (sandy), medium (loamy), heavy (clay), and well-drained soil. It can grow in acidic, neutral, and basic (alkaline) soils, and in semi-shade (light woodland) or no shade. It prefers moist soil.

The fruit stalks of "sweet cherry" are astringent, diuretic, and tonic. A decoction is used in the treatment of cystitis, edema, bronchial complaints, looseness of the bowels, and anemia. An aromatic resin can be obtained by making small incisions in the trunk; this has been used as an inhalant in the treatment of persistent coughs. Although no specific mention has been made of this species, all members of the genus contain amygdalin ($C_{20}H_{27}NO_{11}$) and prunasin ($C_{14}H_{17}NO_6$), substances which break down in water to form HCN (Zargari, 2014; Mozaffarian, 2011; pfaf.org; Mozaffarian, 1996; Henning & Herrmann, 1980).

Ceratonia siliqua (Figure 2.21) is an evergreen tree that grows up to a size of 15 m at a medium rate. It is frost tender. The species is dioecious and pollinated by wasps and flies. It has a preference for light (sandy), medium (loamy), and well-drained soil, and can grow in nutritionally poor soil. It can grow in acidic, neutral, and basic (alkaline) soils, as well as in very alkaline soils. It cannot grow in the shade. It prefers dry or moist soil and can tolerate drought and strong winds but not maritime exposure.

The pulp in the seedpods of "carob" is very nutritious and, due to its high sugar content, sweet tasting and mildly laxative. However, the pulp in the pods is also astringent and, used in a decoction, treats diarrhea and gently helps to cleanse and relieve irritation within the gut. Whilst these appear to be contradictory effects, carob is an example of how the body responds to herbal medicines in different ways, according

FIGURE 2.20 *Cerasus avium* (tree, fruits, and blossoms).

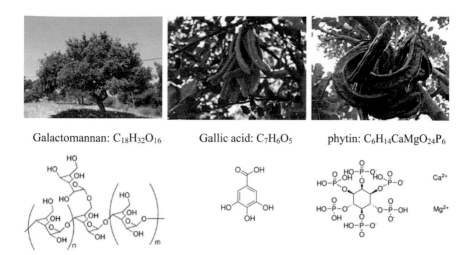

Galactomannan: $C_{18}H_{32}O_{16}$ Gallic acid: $C_7H_6O_5$ phytin: $C_6H_{14}CaMgO_{24}P_6$

FIGURE 2.21   *Ceratonia siliqua* (tree and pods) and its main components: Galactomannan, gallic acid, phytin.

to how the herb is prepared and according to the specific medical problem. The seed-pods are also used in the treatment of coughs. A flour made from the ripe seedpods is demulcent and emollient. It is used in the treatment of diarrhea. The seed husks are astringent and purgative. The bark is strongly astringent (Zargari, 2014; Mozaffarian, 2011; pfaf.org; Mahtout *et al.*, 2018).

Cercis siliquastrum (Figure 2.22) is a deciduous bushy tree or small tree with heart-shaped leaves that grows 10 cm in width and clusters of bright pink pea-flowers open before or with the leaves, followed by flattened, deep purple pods up to 12 cm in length. The flowers are pollinated by bees.

The "Judas tree" causes constipation and is useful for treating bronchitis. The flowers are edible even when raw and are said to have a sweetly acidic taste to them; they can be used as a condiment in salads. The green unripe seedpods are also edible. In Syria the buds are used to make a medicinal tea, used because of its anti-inflammatory properties. The buds are also used as an anti-coagulant, anti-diabetic, anti-malarial, and for skin conditions (Zargari, 2014; Mozaffarian, 2011; pfaf.org; Dayeni & Omidbaigi, 2013).

FIGURE 2.22   *Cercis siliquastrum* (leaves, flowers, and shrub) and its main components: Oleic acid, palmitic acid.

Cheiranthus cheiri (Figure 2.23) is an evergreen perennial that grows up to a size of 0.5 m. The flowers are hermaphrodite and pollinated by bees and flies. It is noted for attracting wildlife. It has a preference for light (sandy), medium (loamy), heavy (clay), and well-drained soil. It can grow in nutritionally poor soil. It prefers neutral and basic (alkaline) soils and can grow in very alkaline soils. It cannot grow in the shade. It prefers dry or moist soil. The plant can tolerate maritime exposure.

"Wallflower" was formerly used mainly as a diuretic and emmenagogue, but recent research has shown that it is more valuable for its effect on the heart. In small doses, it is a cardiotonic, supporting a failing heart in a similar manner to foxglove (*Digitalis purpurea*). In larger doses, however, it is toxic and so is seldom used in herbal medicine. The flowers and stems are anti-rheumatic, anti-spasmodic, cardiotonic, emmenagogue, nervine, purgative, and resolvent. They are used in the treatment of impotence and paralysis. The essential oil is normally used. This should be used with caution because, as mentioned, large doses are toxic. The plant contains the chemical compound cheiranthin which has a stronger cardiotonic action than digitalis (obtained from Digitalis species). This plant should not be used medicinally without expert supervision due to its poisonous nature when taken in large doses. The seeds are aphrodisiac, diuretic, expectorant, stomachic, and tonic. They are used in the treatment of dry bronchitis, fevers, and injuries to the eyes (Zargari, 2014; Mozaffarian, 2011; pfaf.org; Hamedi *et al.*, 2013).

Chimonanthus fragrans (Ch. praecox) (Figure 2.24) is a deciduous bushy tree that grows to a size of 3 m by 3 m at a medium rate. The flowers are hermaphrodite and pollinated by insects. It has a preference for light (sandy), medium (loamy), heavy (clay), and well-drained soil. It can grow in acidic, neutral, and basic (alkaline) soils, as well as in very alkaline soils. It cannot grow in the shade. It prefers moist soil.

The flowers and flower buds of "wintersweet" contain 0.5–0.6% essential oils comprising benzyl alcohol (C_7H_8O), benzyl acetate ($C_6H_5CH_2OCOCH_3$), linalool

Quinoline: C_9H_7N

FIGURE 2.23 *Cheiranthus cheiri* (whole plant and flowers) and its main components: Quinolone, and quercetin (not shown).

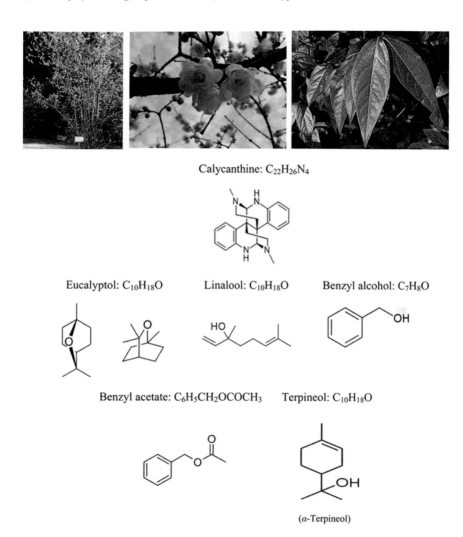

Calycanthine: $C_{22}H_{26}N_4$

Eucalyptol: $C_{10}H_{18}O$ Linalool: $C_{10}H_{18}O$ Benzyl alcohol: C_7H_8O

Benzyl acetate: $C_6H_5CH_2OCOCH_3$ Terpineol: $C_{10}H_{18}O$

(α-Terpineol)

FIGURE 2.24 *Chimonanthus fragrans* (shrub, flowers, and leaves) and its main components: Calycanthine, eucalyptol, linalool, benzyl alcohol, benzyl acetate, terpineol.

($C_{10}H_{18}O$), terpineol ($C_{10}H_{18}O$), and indole (C_8H_7N). They are sialagogue. The flowers are used in the treatment of thirst and depression whilst the essential oil is used to treat colds. The leaves and roots can be used in the treatment of contusions, cuts, hemorrhages, strains, lumbago, rheumatism, numbness, and colds. The wild plants possessed 64 special components such as o-cymene ($C_{10}H_{14}$), eucalyptol ($C_{10}H_{18}O$), γ-terpinene ($C_{10}H_{16}$), linalool ($C_{10}H_{18}O$), etc. (Zargari, 2014; Mozaffarian, 2011; pfaf. org; Mozaffarian, 1996; Juan *et al.*, 2012).

Cicer arietinum (Figure 2.25) is an annual that grows up to a size of 0.6 m. It is not frost tender. The species is hermaphrodite. It has a preference for light (sandy), medium (loamy), and well-drained soil. It can grow in acidic, neutral, and basic (alkaline) soils, as well as in very alkaline soils. It cannot grow in the shade. It prefers dry or moist soil and can tolerate drought.

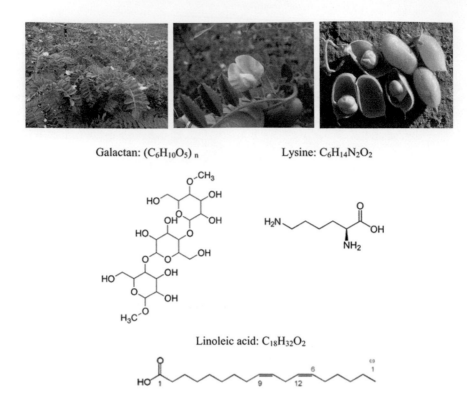

Galactan: $(C_6H_{10}O_5)_n$ Lysine: $C_6H_{14}N_2O_2$

Linoleic acid: $C_{18}H_{32}O_2$

FIGURE 2.25   *Cicer arietinum* (leaves, flower, and pods) and its main components: Galactan, lysine, linoleic acid.

An acid exudation from the seedpods of the "chickpea" is astringent. It has been used in the treatment of dyspepsia, constipation, and snake bites. In addition, it is useful for treating inflammation, cancer, diarrhea, diabetes, and anxiety (Zargari, 2014; Mozaffarian, 2011; pfaf.org; Al Snafi, 2016).

Citrullus colocynthis (Figure 2.26) is an annual or a perennial plant (in the wild). It is native to the Mediterranean basin and Asia. *C. colocynthis* is a desert viny plant that grows in sandy, arid soils. It has a great survival rate under extreme conditions. In fact, it can tolerate annual precipitation of 250 mm to 1500 mm and an annual temperature of 14°C to 27°C. The roots are large, fleshy, and perennial, leading to a high survival rate due to the long tap root. The flowers are yellow and solitary in the axes of leaves and are borne by yellow-greenish peduncles.

"Colocynth" (bitter apple, bitter cucumber) has been widely used in traditional medicine for centuries. In premodern medicine, it was an ingredient in the electuary called confectio hamech, or diacatholicon, and other laxative pills. The fruits are strong aperient, and are also used for treating visceral paralysis (Zargari, 2014; Mozaffarian, 2011; pfaf.org; Uma *et al.*, 2014).

Citrullus vulgaris (Figure 2.27) is an annual that grows to a size of 0.5 m by 2 m. The flowers are monoecious and pollinated by insects. The plant is self-fertile. It has a preference for light (sandy), medium (loamy), and well-drained soil. It can grow in

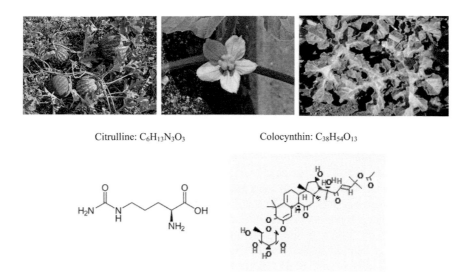

Citrulline: $C_6H_{13}N_3O_3$ Colocynthin: $C_{38}H_{54}O_{13}$

FIGURE 2.26 *Citrullus colocynthis* (fruits, flower, and leaves) and its main components: Citrulline, colocynthin.

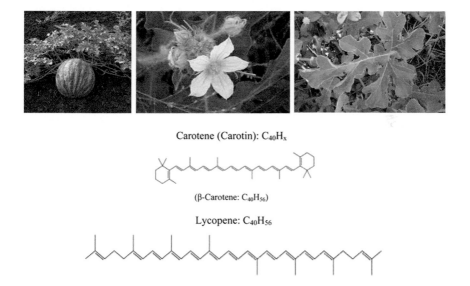

Carotene (Carotin): $C_{40}H_x$

(β-Carotene: $C_{40}H_{56}$)

Lycopene: $C_{40}H_{56}$

FIGURE 2.27 *Citrullus vulgaris* (fruit, flower, and leaves) and its main components: Carotene, lycopene.

acidic, neutral, and basic (alkaline) soils. It cannot grow in the shade. It prefers dry or moist soil and can tolerate drought.

The seed of the "watermelon" is demulcent, diuretic, pectoral, and tonic. It is sometimes used in the treatment of the urinary passages and has been used to treat bedwetting. The seed is also a good vermifuge and has a hypotensive action. A fatty oil in the seed, as well as aqueous or alcoholic extracts, paralyze tapeworms and roundworms. The fruit, eaten when fully ripe or even when almost putrid, is used as a febrifuge. The fruit is also diuretic, being effective in the treatment of dropsy and renal stones. The fruit contains the substance lycopene ($C_{40}H_{56}$). This substance has been shown to protect the body from heart attacks and is more effective when it is cooked. The rind of the fruit is prescribed in cases of alcoholic poisoning and diabetes. The root is purgative and in large doses is said to be a certain emetic. The sprouting seed produces a toxic substance in its embryo (Zargari, 2014; Mozaffarian, 2011; pfaf.org; Akinyele & Oloruntoba, 2017).

Citrus bigaradia (C. aurantium) (Figure 2.28) is an evergreen tree that grows to a size of 9 m by 6 m. The species is hermaphrodite and pollinated by apomixis and insects. The plant is self-fertile. It has a preference for medium (loamy), heavy (clay), and well-drained soil. It can grow in acidic, neutral, and very alkaline soils. It cannot grow in the shade. It prefers moist soil.

It is known as "bitter orange" (sour orange). The citrus species contain a wide range of active ingredients and research is still underway in finding uses for them. They are rich in vitamin C ($C_6H_8O_6$), flavonoids, acids, and volatile oils. They also

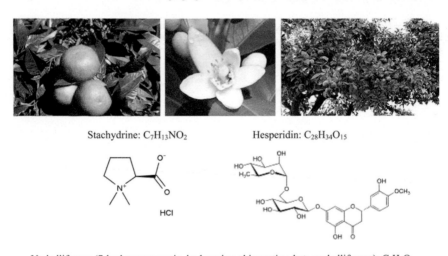

Stachydrine: $C_7H_{13}NO_2$ Hesperidin: $C_{28}H_{34}O_{15}$

Umbelliferone (7-hydroxycoumarin, hydrangine, skimmetine, beta-umbelliferone): $C_9H_6O_3$

Bergapten: $C_{12}H_8O_4$

FIGURE 2.28 *Citrus bigaradia* (fruits, blossom, and tree) and its main components: Stachydrine, hesperidin, umbelliferone, bergapten.

contain coumarins such as bergapten ($C_{12}H_8O_4$) which sensitizes the skin to sunlight. Bergapten ($C_{12}H_8O_4$) is sometimes added to tanning preparations since it promotes pigmentation in the skin, though it can cause dermatitis or allergic responses in some people. Some of the plant's more recent applications are as sources of antioxidants and chemical exfoliants in specialized cosmetics. The plants also contain umbelliferone ($C_9H_6O_3$), which is anti-fungal, as well as essential oils that are anti-fungal and anti-bacterial. Both the leaves and the flowers are anti-spasmodic, digestive, and sedative. An infusion is used in the treatment of stomach problems, sluggish digestion, etc. The fruit is anti-emetic, anti-tussive, carminative, diaphoretic, digestive, and expectorant. The immature fruit can be used, or the mature fruit with seeds and endocarp removed. The immature fruit has a stronger action. They are used in the treatment of dyspepsia, constipation, abdominal distension, stuffy sensation in the chest, prolapse of the uterus, rectum, and stomach. The fruit peel is bitter, digestive, and stomachic. The seed and the pericarp are used in the treatment of anorexia, chest pains, colds, coughs, etc. The essential oil is used in aromatherapy. Its keyword is "radiance." It is used in treating depression, tension, and skin problems due to exposure (Zargari, 2014; Mozaffarian, 2011; pfaf.org; Mozaffarian, 1996; Suntar *et al.*, 2018).

Citrus limetta (Figure 2.29) is a small tree up to 8 m in height, with irregular branches and relatively smooth, brownish-grey bark. It has numerous thorns, which are 1.5–7.5 cm long. The leaves are compound, with acuminate leaflets 5–17 cm long and 2.8–8 cm wide. Its flowers are white, 2–3 cm wide. The fruits are oval and green, ripening to yellow, with greenish pulp. Despite the name "sweet lime," the fruit is greenish-orange in appearance.

Like most citrus, the fruit is rich in vitamin C ($C_6H_8O_6$), providing 50 mg per 100 g serving. In Iran, it is used to treat influenza and the common cold (Zargari, 2014; Mozaffarian, 2011; pfaf.org; Abdullah Khan *et al.*, 2016).

Citrus limonum (C. limon) (Figure 2.30) is an evergreen bushy tree that grows to a size of 3 m by 1 m at a medium rate. It is frost tender. The flowers are hermaphrodite and pollinated by apomixis and insects. The plant is self-fertile. It has a preference for medium (loamy), heavy (clay), and well-drained soil. It can grow in acidic, neutral, and basic (alkaline) soils, as well as in very alkaline soils. It cannot grow in the shade. It prefers moist soil.

"Lemon" is an excellent preventative medicine and has a wide range of uses in the domestic medicine chest. The fruit is rich in vitamin C ($C_6H_8O_6$) which helps the body to fight off infections and also to prevent or treat scurvy. It was at one time a legal requirement that sailors should be given an ounce of lemon each day in order to prevent scurvy. Applied locally, the juice is a good astringent and is used as a gargle

FIGURE 2.29 *Citrus limetta* (fruits, blossoms, and tree).

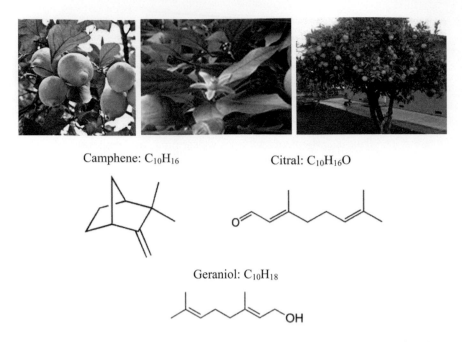

FIGURE 2.30   *Citrus limonum* (fruits, blossom, and tree) and its main components: Camphene, citral, geraniol.

for sore throats, etc. Lemon juice is also a very effective bactericide. It is also a good anti-periodic and has been used in treating malaria and other fevers. Although the fruit is very acidic, once eaten it has an alkalizing effect upon the body. This makes it useful in the treatment of rheumatic conditions. The skin of the ripe fruit is carminative and stomachic. The essential oil from the bark of the fruit is strongly rubefacient and when taken internally in small doses it has stimulating and carminative properties. The stem bark is bitter, stomachic, and tonic. An essential oil from the fruit rind is used in aromatherapy. Its keyword is "refreshing." The citrus species contain a wide range of active ingredients and research is still underway in finding uses for them. They are rich in vitamin C ($C_6H_8O_6$), bioflavonoids, acids, and volatile oils. They also contain coumarins such as bergapten ($C_{12}H_8O_4$) which sensitizes the skin to sunlight. Bergapten ($C_{12}H_8O_4$) is sometimes added to tanning preparations since it promotes pigmentation in the skin, though it can cause dermatitis or allergic responses in some people. Some of the plant's more recent applications are as sources of antioxidants and chemical exfoliants in specialized cosmetics. The bioflavonoids in the fruit help to strengthen the inner lining of blood vessels, especially veins and capillaries, and help counter varicose veins and easy bruising. There is low potential for sensitization through skin contact with volatile oil (Zargari, 2014; Mozaffarian, 2011; pfaf.org; Mozaffarian, 1996; Mathew *et al.*, 2012).

Citrus nobilis (C. reticulata) (Figure 2.31) is an evergreen tree that grows to a size of 4.5 m by 3 m. It is in leaf all year. The species is hermaphrodite and pollinated by apomixis and insects. The plant is self-fertile. It has a preference for medium (loamy), heavy (clay), and well-drained soil. It can grow in acidic, neutral, and basic (alkaline)

Nobiletin: $C_{21}H_{22}O_8$

FIGURE 2.31   *Citrus nobilis* (blossoms, fruits, and tree) and a main component: Nobiletin.

soils, as well as in very acidic and very alkaline soils. It cannot grow in the shade. It prefers moist soil.

The "mandarin orange," the same as all of the citrus species, contains a wide range of active ingredients and research is still underway in finding uses for them. It is rich in vitamin C ($C_6H_8O_6$), flavonoids, acids, and volatile oils. It also contains coumarins such as bergapten ($C_{12}H_8O_4$). Some of the plant's more recent applications are as sources of antioxidants and chemical exfoliants in specialized cosmetics. The fruit is anti-emetic, aphrodisiac, astringent, laxative, and tonic. The flowers are stimulant. The pericarp is analgesic, anti-asthmatic, anti-cholesterolemic, anti-inflammatory, anti-scorbutic, antiseptic, anti-tussive, carminative, expectorant, and stomachic. It is used in the treatment of dyspepsia, gastro-intestinal distension, cough with profuse phlegm, hiccups, and vomiting. The endocarp is carminative and expectorant. It is used in the treatment of dyspepsia, gastro-intestinal distension, coughs, and profuse phlegm. The unripened green exocarp is carminative and stomachic. It is used in the treatment of pain in the chest and hypochondrium, gastro-intestinal distension, swelling of the liver and spleen, and cirrhosis of the liver. The seed is analgesic and carminative. It is used in the treatment of hernias, lumbago, mastitis, and pain or swellings of the testes (Zargari, 2014; Mozaffarian, 2011; pfaf.org; Mozaffarian, 1996; Apraj & Pandita, 2014).

Citrus paradisi (Figure 2.32) is an evergreen tree that grows to a size of 7 m by 7 m at a medium rate. It has a preference for light (sandy), medium (loamy), heavy (clay), and well-drained soil. It can grow in acidic, neutral, and basic (alkaline) soils, as well as in very acidic soils. It cannot grow in the shade.

"Grapefruit" is a rich source of vitamin C ($C_6H_8O_6$), contains the fiber pectin, and the pink and red hues contain the beneficial antioxidant lycopene ($C_{40}H_{56}$). Studies have shown grapefruit helps lower cholesterol, and there is evidence that the seeds have antioxidant properties. Grapefruit forms a core part of the "grapefruit diet," the theory being that the fruit's low glycemic index is able to help the body's metabolism burn fat. It is antioxidant, antidepressant, anti-bacterial, anti-atherogenic,

FIGURE 2.32   *Citrus paradisi* (fruits, blossom, and leaves).

anti-inflammatory, and anti-HIV (Zargari, 2014; Mozaffarian, 2011; pfaf.org; Uckoo *et al.*, 2012).

Citrus sinensis (Figure 2.33) is an evergreen tree that grows up to a size of 9 m. The flowers are hermaphrodite and pollinated by apomixis and insects. The plant is self-fertile. It has a preference for medium (loamy), heavy (clay), and well-drained soil. It can grow in acidic, neutral, and basic (alkaline) soils, as well as in very acidic and very alkaline soils. It cannot grow in the shade. It prefers moist soil.

"Sweet orange" contains a wide range of active ingredients and research is still underway in finding uses for them. They are rich in vitamin C ($C_6H_8O_6$), flavonoids, acids, volatile oils, and bergapten ($C_{12}H_8O_4$). The fruit is appetizer and blood purifier. It is used to allay thirst in people with fevers and also treats catarrh. The fruit juice is useful in the treatment of bilious affections and bilious diarrhea. The fruit rind is carminative and tonic. The fresh rind is rubbed on the face as a treatment for acne. The dried peel is used in the treatment of anorexia, colds, coughs, etc. (Zargari, 2014; Mozaffarian, 2011; pfaf.org; Senthamil Selvi *et al.*, 2016).

Consolida regalis (Figure 2.34) is an annual/biennial that grows up to a size of 0.5 m. The flowers are hermaphrodite and pollinated by bees, moths, and butterflies. The plant is self-fertile. It is noted for attracting wildlife. It has a preference for light (sandy), medium (loamy), heavy (clay), and well-drained soil. It can grow in acidic, neutral, and basic (alkaline) soils, and semi-shade (light woodland) or no shade. It prefers dry or moist soil and can tolerate drought.

"Larkspur" was at one time used internally in the treatment of a range of diseases, but its only certain action is as a violent purgative and nowadays it is only occasionally

FIGURE 2.33   *Citrus sinensis* (fruits, blossoms, and tree).

Delsoline (Acominine): $C_{25}H_{41}NO_7$

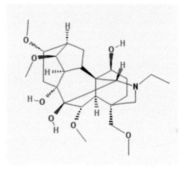

FIGURE 2.34 *Consolida regalis* (flowers and whole plant) and its main components: Delsoline, and kaempferol (not shown).

used in folk medicine. It is of value, however, when used externally, to kill skin parasites. The plant should be used with caution (see the notes above on toxicity). The seed is anthelmintic, mildly diuretic, hypnotic, purgative, and vasodilator. It has been used internally in the treatment of spasmodic asthma and dropsy. The flowers or the whole plant are mildly diuretic and hypotensive. The expressed juice of the leaves has been considered an effective application to bleeding piles. A conserve made from the flowers has been seen as a good remedy for children when subject to violent purging. The juice of the flowers has also been used as a treatment for colic. All parts of the plant are poisonous in large doses. The seed is especially toxic (Zargari, 2014; Mozaffarian, 2011; pfaf.org; Strzelecka, 1965).

Coriandrum sativum (Figure 2.35) is an annual that grows to a size of 0.5 m by 0.3 m. It is not frost tender. The flowers are hermaphrodite and pollinated by insects. The plant is self-fertile. It is noted for attracting wildlife. It has a preference for light (sandy), medium (loamy), and well-drained soil. It can grow in acidic, neutral, and basic (alkaline) soils, as well as in very alkaline soils, semi-shade (light woodland), or no shade. It prefers dry or moist soil.

"Coriander" is a commonly used domestic remedy, valued especially for its effect on the digestive system, treating flatulence, diarrhea, and colic. It settles spasms in the gut and counters the effects of nervous tension. The seed is aromatic, carminative, expectorant, narcotic, stimulant, and stomachic. It is most often used with active purgatives in order to disguise their flavor and combat their tendency to cause gripe. The raw seed is chewed to stimulate the flow of gastric juices and to eliminate foul breath

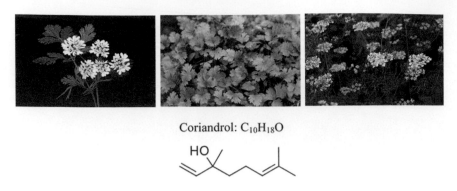

Coriandrol: $C_{10}H_{18}O$

FIGURE 2.35 *Coriandrum sativum* (flowers, leaves, and whole plant) and its main components: Coriandrol, and phellandrene (not shown).

and will sweeten the breath after garlic has been eaten. Some caution is advised, however, because if used too freely the seeds become narcotic. Externally the seeds have been used as a lotion or have been bruised and used as a poultice to treat rheumatic pains. The essential oil is used in aromatherapy. Its keyword is "appetite stimulant." The plant can have a narcotic effect if it is eaten in very large quantities. Powdered coriander and oil may cause allergic reactions and photosensitivity. Use dry coriander sparingly if suffering from bronchial asthma and chronic bronchitis (Zargari, 2014; Mozaffarian, 2011; pfaf.org; Al Snafi, 2016).

    Cornus mas (Figure 2.36) is a deciduous bushy tree that grows to a size of 5 m by 5 m at a medium rate. The species is hermaphrodite and pollinated by bees. It has a preference for light (sandy) and medium (loamy), and can grow in heavy clay soils. Also, it can grow in acidic, neutral, and basic (alkaline) soils, as well as in very alkaline soils and semi-shade (light woodland) or no shade. It prefers moist soil. The plant can tolerate strong winds but not maritime exposure.

    The raw, dried fruit of "cornelian cherry" is used in preserves with a juicy, nice, acidic flavor. The fully ripe fruit has a somewhat plum-like flavor and texture and is very nice to eat, but the unripe fruit is rather astringent. It is rather low in pectin and

Glyoxylic acid (Oxoacetic acid): $C_2H_2O_3$

FIGURE 2.36 *Cornus mas* (tree, fruits, and leaves) and its main component: Glyoxylic acid.

so needs to be used with other fruit when making jam. At one time the fruit was kept in brine and used like olives. The fruit is a reasonable size, up to 15 mm long, with a single large seed. A small amount of edible oil can be extracted from the seeds. The seeds can be roasted, ground into a powder, and used as a coffee substitute. The bark and the fruit are astringent, febrifuge, and nutritive. The astringent fruit is a good treatment for bowel complaints and fevers, whilst it is also used in the treatment of cholera. The flowers are used in the treatment of diarrhea (Zargari, 2014; Mozaffarian, 2011; pfaf.org; Hosseinpour Jaghdani *et al.*, 2017).

Cornus sanguinea (Figure 2.37) is a deciduous bushy tree that grows up to a size of 3 m. The species is hermaphrodite and pollinated by insects. It has a preference for light (sandy) and medium (loamy), and can grow in heavy clay soil. It can grow in acidic, neutral, basic (alkaline), and very alkaline soils, as well as in semi-shade (light woodland) or no shade. It prefers moist soil.

The bark of "common dogwood" (bloody dogwood) is astringent and febrifuge. It is used to treat fever. The leaves are sometimes used externally as an astringent. The fruit is emetic (Zargari, 2014; Mozaffarian, 2011; pfaf.org; Zorica *et al.*, 2017).

Cucumis melo (Figure 2.38) is an annual climber that grows to a size of 1.5 m. The flowers are monoecious and pollinated by insects. The plant is self-fertile. It has a preference for light (sandy), medium (loamy), heavy (clay), and well-drained soil. It

FIGURE 2.37 *Cornus sanguinea* (shrub, leaves, and fruits).

Cellulose: $(C_6H_{10}O_5)_n$

FIGURE 2.38 *Cucumis melo* (flowers, fruit, and leaves) and its main components: Cellulose, and vitamin C (not shown).

can grow in acidic, neutral, and basic (alkaline) soils. It cannot grow in the shade. It prefers moist soil.

The fruit of the "Persian melon" can be used as a cooling light cleanser or moisturizer for the skin. It is also used as a first aid treatment for burns and abrasions. The flowers are expectorant and emetic. The fruit is stomachic. The seed is anti-tussive, digestive, febrifuge, and vermifuge. When used as a vermifuge, the whole seed complete with the seed coat is ground into a fine flour, then made into an emulsion with water and eaten. It is then necessary to purge in order to expel the tapeworms or other parasites from the body. The root is diuretic and emetic. The sprouting seed produces a toxic substance in its embryo (Zargari, 2014; Mozaffarian, 2011; pfaf.org; Al Haidari *et al.*, 2010).

Cucumis sativus (Figure 2.39) is an annual climber that grows to a size of 2 m. The flowers are monoecious and pollinated by insects. The plant is self-fertile. It has a preference for light (sandy), medium (loamy), heavy (clay), and well-drained soil. It can grow in acidic, neutral, and basic (alkaline) soils. It cannot grow in the shade. It prefers moist soil.

The leaf juice of the "cucumber" is emetic. It is used to treat dyspepsia in children. The fruit is depurative, diuretic, emollient, purgative, and resolvent. The fresh fruit is used internally in the treatment of blemished skin, heat rash, etc., whilst it is used externally as a poultice for burns, sores, etc., and as a cosmetic for softening the skin. The seed is cooling, diuretic, tonic, and vermifuge. 25–50 grams of the thoroughly ground seeds (including the seed coat) is a standard dose as a vermifuge and usually needs to be followed by a purgative to expel the worms from the body. A decoction of the root is diuretic. The sprouting seed produces a toxic substance in its embryo (Zargari, 2014; Mozaffarian, 2011; pfaf.org; Sebastian *et al.*, 2010).

Cuminum cyminum (Figure 2.40) is an annual that grows to a size of 0.3 m by 0.2 m at a fast rate. The species is hermaphroditic and pollinated by insects. The plant is self-fertile. It is noted for attracting wildlife. It has a preference for light (sandy), medium (loamy), heavy (clay), and well-drained soil. It can grow in acidic, neutral, and basic (alkaline) soils, as well as in very acidic and very alkaline soils. It cannot grow in the shade. It prefers dry or moist soil and can tolerate drought.

"Cumin" is an aromatic, astringent herb that benefits the digestive system and acts as a stimulant to the sexual organs. It has been used in the treatment of minor digestive complaints, chest conditions, and coughs, and as a painkiller and to treat rotten teeth. Cumin is seldom used in Western herbal medicine, having been superseded by caraway which has similar properties but a more pleasant flavor. It is still widely used in India, however, where it is said to promote the assimilation of other herbs and to improve liver function. The seed is anti-spasmodic, carminative, galactagogue,

FIGURE 2.39   *Cucumis sativus* (fruits, flower, and leaves).

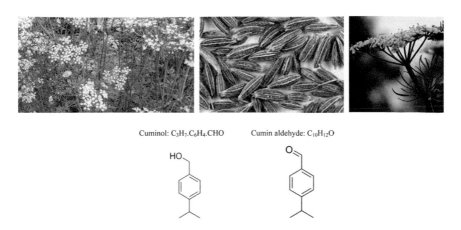

Cuminol: $C_3H_7.C_6H_4.CHO$ Cumin aldehyde: $C_{10}H_{12}O$

FIGURE 2.40 *Cuminum cyminum* (whole plant, seeds, and flower) and main components: Cuminol, cumin aldehyde, o-cymene.

stimulant, and stomachic. A general tonic to the whole digestive system, it is used in the treatment of flatulence and bloating, reducing intestinal gas and relaxing the gut as a whole. In India it is also used in the treatment of insomnia, colds, and fevers, and to improve milk production in nursing mothers. Ground into a powder and mixed into a paste with onion juice, it can be applied to scorpion stings. The herb has been used externally as a poultice to relieve a stitch and pains in the side. The essential oil obtained from the seed is anti-bacterial and larvicidal (Zargari, 2014; Mozaffarian, 2011; pfaf.org; Dhaliwali *et al.*, 2016).

Cydonia oblonga (Figure 2.41) is a deciduous tree that grows to a size of 7.5 m by 7 m at a medium rate. The species is hermaphrodite and pollinated by insects. It has a preference for light (sandy), medium (loamy), and heavy (clay) soils. It can grow in

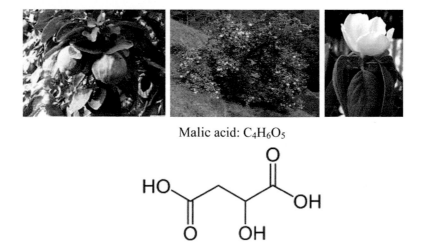

Malic acid: $C_4H_6O_5$

FIGURE 2.41 *Cydonia oblonga* (fruits, tree, and flower) and its main components: Malic acid, and amygdalin (not shown).

acidic, neutral, and basic (alkaline) soils, and in full shade (deep woodland), semi-shade (light woodland), or no shade. It prefers moist soil.

The stem bark of "quince" is astringent and is used in the treatment of ulcers. The seed is a mild but reliable laxative, astringent, and anti-inflammatory. When soaked in water, the seed swells up to form a mucilaginous mass. This has a soothing and demulcent action when taken internally and is used in the treatment of respiratory diseases, especially in children. This mucilage is also applied externally to minor burns, etc. The fruit is anti-vinous, astringent, cardiac, carminative, digestive, diuretic, emollient, expectorant, pectoral, peptic, refrigerant, restorative, stimulant, and tonic. The unripe fruit is very astringent; a syrup made from it is used in the treatment of diarrhea and is particularly safe for children. The fruit, and its juice, can be used as a mouthwash or can be gargled to treat mouth ulcers, gum problems, and sore throats. The leaves contain tannin and pectin (Zargari, 2014; Mozaffarian, 2011; pfaf.org; Ashraf *et al.*, 2016).

Daucus carota (Figure 2.42) is a biennial growing to 0.6 m by 0.3 m at a medium rate. It is not frost tender. The flowers are hermaphrodite and pollinated by flies and beetles. The plant is self-fertile. It is noted for attracting wildlife. It has a preference for light (sandy), medium (loamy), heavy (clay), and well-drained soil. It can grow in acidic, neutral, and basic (alkaline) soils. It cannot grow in the shade. It prefers moist soil. The plant can tolerate maritime exposure.

"Wild carrot" is an aromatic herb that acts as a diuretic, soothes the digestive tract, and stimulates the uterus. A wonderfully cleansing medicine, it supports the liver, stimulates the flow of urine, and the removal of waste by the kidneys. The whole plant is anthelmintic, carminative, deobstruent, diuretic, galactogogue, ophthalmic, and stimulant. An infusion is used in the treatment of various complaints including digestive disorders, kidney and bladder diseases, and in the treatment of dropsy. An infusion of the leaves has been used to counter cystitis and kidney stone formation, and to diminish stones that have already formed. Carrot leaves contain significant amounts of porphyrins, which stimulate the pituitary gland and lead to the release of increased levels of sex hormones. The plant is harvested in July and dried for later use. A warm water infusion of the flowers has been used in the treatment of diabetes. The grated raw root, especially of the cultivated forms, is used as a remedy for threadworms. The root is also used to encourage delayed menstruation. The root of the wild plant can induce uterine contractions and so should not be used by pregnant women. A tea made from the roots is diuretic and has been used in the treatment of urinary stones. The seeds are diuretic, carminative, emmenagogue, and anthelmintic. An infusion is used in the treatment of edema, flatulent indigestion, and menstrual problems. The seed is a traditional "morning after" contraceptive and there is some evidence to uphold this belief; it requires further investigation. Carrot seeds can be abortifacient and so should not be used by pregnant women. Carrots sometimes cause allergic reactions in some people. Skin contact with the sap is said to cause photosensitivity and/or dermatitis in some people. Daucus has been reported to contain asarone ($C_{12}H_{16}O_3$), choline ($C_5H_{14}NO$), ethanol (C_2H_5OH), formic acid (CH_2O_2), hydrocyanic acid (HCN), butyric acid ($C_4H_8O_2$), limonene ($C_{10}H_{16}$), malic acid ($C_4H_6O_5$), maltose ($C_{12}H_{22}O_{11}$), oxalic acid ($C_2H_2O_4$), palmitic acid ($C_{16}H_{32}O_2$), pyrrolidine (C_4H_9N), and quinic acid ($C_7H_{12}O_6$). Handling carrot foliage, especially wet foliage, can cause irritation and vesication. Sensitized photosensitive persons may get an exact reproduction of the leaf on the skin by placing the leaf on the skin for a while followed by exposure to sunshine (Zargari, 2014; Mozaffarian, 2011; pfaf.org; Ayeni *et al.*, 2018).

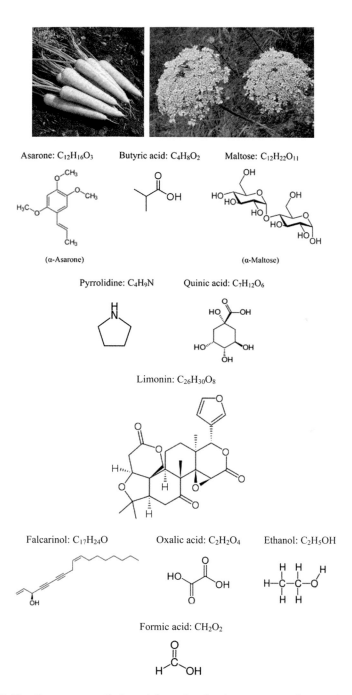

FIGURE 2.42 *Daucus carota* (fruits and flowers) and main components: Asarone, butyric acid, maltose, pyrrolidine, quinic acid, limonin, falcarinol, oxalic acid, ethanol, formic acid.

Descurainia sophia (Figure 2.43) is an annual/biannual that grows up to a size of 0.9 m. The flowers are hermaphrodite and self-pollinated. The plant is self-fertile. It has a preference for light (sandy), medium (loamy), and heavy (clay) soils. It can grow in acidic, neutral, and basic (alkaline) soils, as well as in semi-shade (light woodland) or no shade. It prefers moist soil.

A poultice of the "flixweed" (herb-Sophia, tansy mustard) has been used to ease the pain of toothache. The juice of the plant has been used in the treatment of chronic coughs, hoarseness, and ulcerated sore throats. A strong decoction of the plant has proved excellent in the treatment of asthma. The flowers and leaves are anti-scorbutic and astringent. The seed is considered to be cardiotonic, demulcent, diuretic, expectorant, febrifuge, laxative, restorative, and tonic. It is used in the treatment of asthma, fevers, bronchitis, edema, and dysentery. It is also used in the treatment of worms and calculus complaints. It is decocted with other herbs for treating various ailments. The seeds have formed a special remedy for sciatica. A poultice of the ground up seeds has been used on burns and sores (Zargari, 2014; Mozaffarian, 2011; pfaf.org; Nimrouzi & Zarshenas, 2016).

Dorema ammoniacum (Figure 2.44) is a perennial that grows to a size of 2.5 m by 1 m. The species is hermaphrodite and pollinated by insects. The plant is self-fertile.

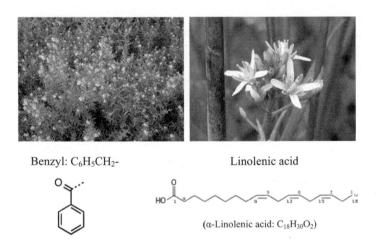

Benzyl: $C_6H_5CH_2$-   Linolenic acid

(α-Linolenic acid: $C_{18}H_{30}O_2$)

FIGURE 2.43 *Descurainia sophia* (whole plant and flowers) and main components: Benzyl, linolenic acid.

FIGURE 2.44 *Dorema ammoniacum* (flowers and whole plant).

It has a preference for light (sandy), medium (loamy), and heavy (clay) soils. It can grow in acidic, neutral, and basic (alkaline) soils, as well as in semi-shade (light woodland) or no shade. It prefers dry or moist soil.

"Ammoniacum" has been used in Western herbal medicine for thousands of years and is still seen as an effective remedy for various complaints of the chest. A gum resin is found in cavities in the tissues of stems, roots, and petioles. It often exudes naturally from holes in the stems caused by beetles though this is not so pure as that obtained from the plant tissues. The resin is anti-spasmodic, carminative, diaphoretic, mildly diuretic, expectorant, poultice, stimulant, and vasodilator. It is often used internally in the treatment of chronic bronchitis (especially in the elderly), asthma, and catarrh. Externally, it is used as a plaster for swellings of the joints and indolent tumors. The resin exudes as a milky gum from holes made in the stems. This gum is pressed into blocks and then ground into a powder (Zargari, 2014; Mozaffarian, 2011; pfaf.org; Zandpour *et al.*, 2016).

Eucalyptus globulus (Figure 2.45) is an evergreen tree that grows to a size of 55 m by 15 m at a fast rate. The flowers are hermaphrodite and pollinated by bees. It has a preference for light (sandy), medium (loamy), heavy (clay), and well-drained soil, and can grow in nutritionally poor soil. It can grow in acidic, neutral, and basic (alkaline) soils. It cannot grow in the shade. It prefers dry moist or wet soil and can tolerate drought.

"Eucalyptus" leaves are a traditional Aboriginal herbal remedy. The essential oil found in the leaves is a powerful antiseptic and is used all over the world for relieving coughs and colds, sore throats, and other infections. The essential oil is a common ingredient in many over-the-counter cold remedies. The adult leaves, without their petioles, are anti-periodic, antiseptic, aromatic, deodorant, expectorant, febrifuge, hypoglycemic, and stimulant. The leaves, and the essential oil they contain,

Citronellal: $C_{10}H_{18}O$

FIGURE 2.45 *Eucalyptus globulus* (leaves, flowers, and fruits) and its main components: Citronellal, and eucalyptol (not shown).

are antiseptic, anti-spasmodic, expectorant, febrifuge, and stimulant. Extracts of the leaves have anti-bacterial activity. The essential oil obtained from various species of eucalyptus is a very powerful antiseptic, especially when it is old, because ozone is formed in it on exposure to air. It has a decided disinfectant action, destroying the lower forms of life. The oil can be used externally, applied to cuts, skin infections, etc. It can be inhaled for treating blocked nasal passages, gargled for a sore throat, and can also be taken internally for a wide range of complaints. Some caution is advised, however, because like all essential oils, it can have a deleterious effect on the body in larger doses. The oil from this species has a somewhat disagreeable odor and so it is no longer used so frequently for medicinal purposes, with other members of the genus being used instead. An oleoresin is exuded from the tree, and can also be obtained from the tree by making incisions in the trunk. This resin contains tannins and is powerfully astringent; it is used internally in the treatment of diarrhea and bladder inflammation, and externally it is applied to cuts, etc. The essential oil is used in aromatherapy. Citronellal, an essential oil found in most Eucalyptus species, is reported to be mutagenic when used in isolation. In large doses, oil of eucalyptus, like so many essential oils, has caused fatalities from intestinal irritation. Death is reported from ingestion of 4–24 ml of essential oils, but recoveries are also reported for the same amount. Symptoms include gastroenteric burning and irritation, nausea, vomiting, diarrhea, oxygen deficiency, weakness, dizziness, stupor, difficult respiration, delirium, paralysis, convulsions, and death, usually due to respiratory failure. The plant is reported to cause contact dermatitis. Sensitive persons may develop urticaria from handling the foliage and other parts of the plant. Avoid if on treatment for diabetes mellitus. Infants and small children—avoid oil preparations on faces as possible life-threatening spasms (Zargari, 2014; Mozaffarian, 2011; pfaf.org; Vecchio *et al.*, 2016).

Faba vulgaris (Figure 2.46) is a fast-growing annual reaching up to 1 m in size. The flowers are hermaphrodite and pollinated by bees. The plant is self-fertile. It can fix nitrogen. It has a preference for light (sandy), medium (loamy), and well-drained soil, and can grow in heavy clay soil. It can grow in acidic, neutral, and basic (alkaline) soils, as well as in semi-shade (light woodland) or no shade. It prefers moist soil. The plant can tolerate strong winds but not maritime exposure.

The "broad bean" is tonic and anti-paroxysm and is useful for treating fatigue. Although often used as an edible seed, there is a report that eating the seed of this plant can cause the favism disease in susceptible people. Favism only occurs in cases of excessive consumption of the seed (no more details are given) and when the person is genetically inclined towards the disease (Zargari, 2014; Mozaffarian, 2011; pfaf.org; Singh *et al.*, 2013).

FIGURE 2.46   *Faba vulgaris* (pods, flowers, and whole plant).

Ferula gummosa (Figure 2.47) is a perennial that grows up to a size of 1 m. The flowers are hermaphrodite and pollinated by flies. The plant is self-fertile. It has a preference for light (sandy), medium (loamy), and heavy (clay) soils, as well as well-drained soil. It can grow in acidic, neutral, and basic (alkaline) soils. It cannot grow in the shade. It prefers dry or moist soil.

The whole plant, but especially the root, contains the gum resin "galbanum," which is anti-spasmodic, carminative, expectorant, and stimulant. It is used internally in the treatment of chronic bronchitis, asthma, and other chest complaints. It is a digestive stimulant and anti-spasmodic, reducing flatulence, griping pains, and colic. Externally it is used as a plaster for inflammatory swellings, ulcers, boils, wounds, and skin complaints (Zargari, 2014; Mozaffarian, 2011; pfaf.org; Saadat talab *et al.*, 2017).

Foeniculum vulgare (Figure 2.48) is an evergreen perennial that grows to a size of 1.5 m by 1 m. It is not frost tender. The flowers are hermaphrodite and pollinated by insects. The plant is self-fertile. It is noted for attracting wildlife. It has a preference for light (sandy), medium (loamy), heavy (clay), and well-drained soil. It can grow in acidic, neutral, and basic (alkaline) soils. It cannot grow in the shade. It prefers dry or moist soil and can tolerate drought. The plant can tolerate strong winds but not maritime exposure.

"Fennel" has a long history of herbal use and is a commonly used household remedy, being useful in the treatment of a variety of complaints, especially those of the digestive system. The seeds, leaves, and roots can be used, but the seeds are most active medicinally and are the part normally used. An essential oil is often extracted from the fully ripened and dried seed for medicinal use, though it should not be given to pregnant women. The plant is analgesic, anti-inflammatory, anti-spasmodic, aromatic, carminative, diuretic, emmenagogue, expectorant, galactogogue, hallucinogenic, laxative, stimulant, and stomachic. An infusion is used in the treatment of indigestion, abdominal distension, stomach pains, etc. It helps in the treatment of

Cadinene: $C_{15}H_{24}$ Cadinol: $C_{15}H_{26}O$

FIGURE 2.47 *Ferula gummosa* (flowers and whole plant) and its main components: Cadinene, cadinol, and umbelliferone (not shown).

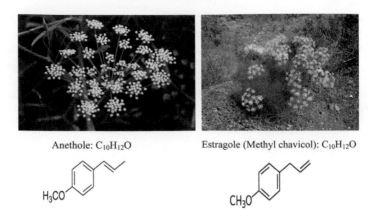

Anethole: $C_{10}H_{12}O$ Estragole (Methyl chavicol): $C_{10}H_{12}O$

FIGURE 2.48   *Foeniculum vulgare* (flowers and whole plant) and its main components: Anethole, estragole, and phellandrene (not shown).

kidney stones and, when combined with a urinary disinfectant, makes an effective treatment for cystitis. It can also be used as a gargle for sore throats and as an eye-wash for sore eyes and conjunctivitis. Fennel is often added to purgatives in order to allay their tendency to cause gripe, and also to improve the flavor. An infusion of the seeds is a safe and effective treatment for wind in babies. An infusion of the root is used to treat urinary disorders. An essential oil obtained from the seed is used in aromatherapy. Its keyword is "normalizing." The essential oil is bactericidal, carminative, and stimulant. Some caution is advised (see earlier notes on toxicity). The German Commission E Monographs: Therapeutic guide to herbal medicines (Blumenthal *et al.*, 1998), approve *Foeniculum vulgare* for coughs, bronchitis, and dyspeptic complaints. Skin contact with the sap or essential oil is said to cause photosensitivity and/or dermatitis in some people. Ingestion of the oil can cause vomiting, seizures, and pulmonary edema. People with cirrhosis/liver disorders should avoid its use and it should not be given to small children. Diabetics should check the sugar content when preparing (Zargari, 2014; Mozaffarian, 2011; pfaf.org; Kocheki Shahmokhtar & Armand, 2017).

Fragaria vesca (Figure 2.49) is a perennial that grows up to a size of 0.3 m. It is not frost tender. The flowers are hermaphrodite and pollinated by insects. It has a preference for light (sandy), medium (loamy), heavy (clay), and well-drained soil. It can grow in acidic, neutral, and basic (alkaline) soils, as well as in semi-shade (light woodland) or no shade. It prefers moist soil.

The leaves and the fruit of the "wild strawberry" are astringent, diuretic, laxative, and tonic. The leaves are mainly used, though the fruits are an excellent food to take when feverish and are also effective in treating rheumatic gout. A slice of strawberry is also excellent when applied externally to sunburnt skin. A tea made from the leaves is a blood tonic. It is used in the treatment of chilblains and also as an external wash on sunburn. The leaves are harvested in the summer and dried for later use. The fruits contain salicylic acid ($C_7H_6O_3$) and are beneficial in the treatment of liver and kidney complaints, as well as in the treatment of rheumatism and gout. The roots are astringent and diuretic. A decoction is used internally in the treatment of diarrhea and chronic dysentery. Externally it is used to treat chilblains and as a throat gargle. The

Salicylic acid: $C_7H_6O_3$

FIGURE 2.49 *Fragaria vesca* (fruits, flower, and leaf) and its main component: Salicylic acid.

roots are harvested in the autumn and dried for later use (Zargari, 2014; Mozaffarian, 2011; pfaf.org; Liberal *et al.*, 2014).

Frangula alnus (Figure 2.50) is a deciduous bushy tree that grows to a size of 5 m by 4 m at a slow rate. The species is hermaphrodite and pollinated by insects. It is noted for attracting wildlife. It has a preference for light (sandy), medium (loamy), and heavy (clay) soils. It can grow in acidic, neutral, and basic (alkaline) soils, as well as in semi-shade (light woodland) or no shade. It prefers moist or wet soil.

"Alder buckthorn" has been used medicinally as a gentle laxative since at least the Middle Ages. The bark contains 3–7% anthraquinone ($C_{14}H_8O_2$), which acts on the wall of the colon stimulating a bowel movement approximately 8–12 hours after ingestion. It is so gentle and effective a treatment when prescribed in the correct dosages that it is completely safe to use for children and pregnant women. The bark also contains anthrone ($C_{14}H_{10}O$) and anthranol ($C_{14}H_{10}O$), these induce vomiting, but the severity of their effect is greatly reduced after the bark has been dried and stored for a long time. The bark is harvested in early summer from the young trunk and moderately sized branches; it must then be dried and stored for at least 12 months before being used. The inner bark is cathartic, cholagogue, laxative (the fresh bark is violently purgative), tonic, and vermifuge. It is taken internally as a laxative for chronic atonic constipation and is also used to treat abdominal bloating, hepatitis, cirrhosis, jaundice, and liver and gall bladder complaints (Zargari, 2014; Mozaffarian, 2011; pfaf.org; Kremer *et al.*, 2011).

Fumaria parviflora (Figure 2.51) is an annual that grows up to a size of 0.3 m. The flowers are hermaphrodite and pollinated by bees and flies. The plant is self-fertile. It has a preference for light (sandy) and medium (loamy) soils. It can grow in acidic, neutral, and basic (alkaline) soils, as well as in semi-shade (light woodland) or no shade. It prefers moist soil.

"Fine-leaf fumitory" (fine-leaved fumitory, Indian fumitory) has been highly valued for its tonic and blood cleansing effect upon the body. It is sweaty and appetizer.

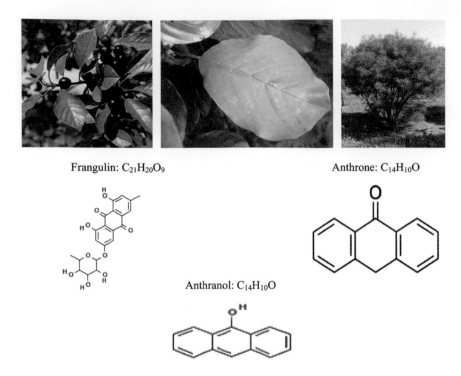

Frangulin: $C_{21}H_{20}O_9$ Anthrone: $C_{14}H_{10}O$

Anthranol: $C_{14}H_{10}O$

FIGURE 2.50 *Frangula alnus* (fruits, leaf, and shrub) and its main components: Frangulin, anthrone, anthranol, and anthraquinone (not shown).

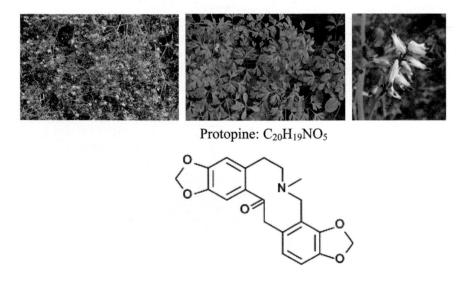

Protopine: $C_{20}H_{19}NO_5$

FIGURE 2.51 *Fumaria parviflora* (whole plant, leaves, and flowers) and its main component: Protopine.

There is protopine ($C_{20}H_{19}NO_5$) in this plant (Zargari, 2014; Mozaffarian, 2011; pfaf.org; Jameel *et al.*, 2014).

Geranium robertianum (Figure 2.52) is an annual/biennial that grows to a size of 0.4 m by 0.4 m. The flowers are hermaphrodite and pollinated by insects. The plant is self-fertile. It has a preference for light (sandy), medium (loamy), heavy (clay), and well-drained soil. It can grow in acidic, neutral, and basic (alkaline) soils and semi-shade (light woodland) or no shade. It prefers dry or moist soil.

"Herb Robert" is little used in modern herbalism, but is occasionally employed as an astringent to halt bleeding, treat diarrhea, etc., in much the same way as *G. maculatum*. The leaves are anti-rheumatic, astringent, mildly diuretic, and vulner-ary. Modern research has shown that the leaves can lower blood sugar levels and so it can be useful in the treatment of diabetes. An infusion of the leaves is used in the treatment of bleeding, stomach ailments, kidney infections, jaundice, etc. Externally, a wash or poultice is applied to swollen and painful breasts, rheumatic joints, bruises, bleeding, etc. It is best to use the entire plant, including the roots. The plant can be harvested at any time from late spring to early autumn and is usually used fresh. A homeopathic remedy is made from the plant. Details of uses are not given in this report (Zargari, 2014; Mozaffarian, 2011; pfaf.org; Catarino *et al.*, 2017).

Glycyrrhiza globra (Figure 2.53) is a perennial that grows to a size of 1.2 m by 1 m. The flowers are hermaphrodite and pollinated by insects. It can fix nitrogen. It has a preference for light (sandy) and medium (loamy) soils. It can grow in acidic, neutral, and basic (alkaline) soils, as well as in semi-shade (light woodland) or no shade. It prefers moist soil. The plant can tolerate strong winds but not maritime exposure.

"Licorice" is one of the most commonly used herbs in Western herbal medicine and has a very long history of use, both as a medicine and also as a flavoring to dis-guise the unpleasant flavor of other medications. It is a very sweet, moist, soothing herb that detoxifies and protects the liver and is also powerfully anti-inflammatory, being used in conditions as varied as arthritis and mouth ulcers. The root is alterative,

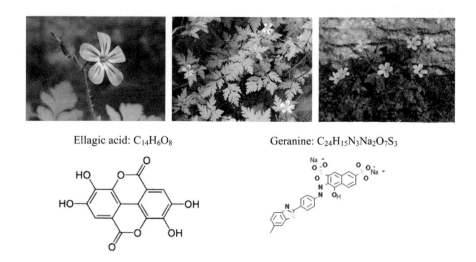

Ellagic acid: $C_{14}H_6O_8$　　　　　　Geranine: $C_{24}H_{15}N_3Na_2O_7S_3$

FIGURE 2.52 *Geranium robertianum* (flower, leaves, and whole plant) and its main components: Ellagic acid, geranine.

Glycyrrhizin (Glycyrrhizic acid): $C_{42}H_{62}O_{16}$

FIGURE 2.53 *Glycyrrhiza glabra* (pods, leaves, and whole plant) and its main component: Glycyrrhizin.

anti-spasmodic, demulcent, diuretic, emollient, expectorant, laxative, moderately pectoral, and tonic. The root has also been shown to have a hormonal effect similar to the ovarian hormone. Licorice root is much used in cough medicines and also in the treatment of catarrhal infections of the urinary tract. It is taken internally in the treatment of Addison's disease, asthma, bronchitis, coughs, peptic ulcer, arthritis, allergic complaints and following steroidal therapy. It should be used in moderation and should not be prescribed for pregnant women or people with high blood pressure, kidney disease, or people taking digoxin-based medication. Prolonged usage raises blood pressure and causes water retention Externally, the root is used in the treatment of herpes, eczema, and shingles. The root is harvested in the autumn when 3–4 years old and is dried for later use. The German Commission E Monographs: Therapeutic guide to herbal medicines (Blumenthal *et al.*, 1998), approve *Glycyrrhiza glabra* for coughs, bronchitis, gastritis. A gross overdose of the root can cause edema, high blood pressure, and congestive heart failure. Do not use during premenstrual syndrome as water retention and bloating occur. Use with caution if pregnant or have liver cirrhosis. Avoid using for more than 6 weeks. Excessive quantities may cause headache, sluggishness, and potassium depletion (Zargari, 2014; Mozaffarian, 2011; pfaf.org; Thakur & Raj, 2017).

Gossypium herbaceum (Figure 2.54) is a shrub that grows up to a size of 1.2 m. It has a preference for light (sandy), medium (loamy), and well-drained soil. It can grow in acidic, neutral, basic (alkaline), and saline soils, as well as in semi-shade (light woodland) or no shade. It prefers moist soil. The plant is not wind tolerant.

"Cotton" is an astringent, slightly acidic, aromatic herb that causes uterine contractions, depresses sperm production, lowers fever, reduces inflammation, and soothes irritated tissue. It also has anti-viral and anti-bacterial actions. The root bark contains gossypol and flavonoids. It is seldom used in modern herbalism, but has been used as a milder and safer alternative to ergot (*Claviceps purpurea*) for inducing uterine contractions in order to speed up a difficult labor. It can induce

β-sitosterol: $C_{29}H_{50}O$

Gossypol: $C_{30}H_{30}O_8$

FIGURE 2.54 *Gossypium herbaceum* (cotton fibers, whole plant, and flower) and its main components: β-sitosterol, gossypol, and choline (not shown).

an abortion or the onset of a period, and reduces total menstrual flow. It has also been taken internally in the treatment of painful menstruation. The root bark also encourages an increased milk flow in nursing mothers and blood clotting. The roots are harvested at the end of the growing season, peeled, and dried. The seeds are taken internally in the treatment of dysentery, intermittent fever, and fibroids. Externally, the seeds are used to treat herpes, scabies, wounds, and orchitis. The oil obtained from the seed contains a substance known as gossypol ($C_{30}H_{30}O_8$), which has the effect of lowering sperm production and possibly causing infertility in males. Research has been carried out into its potential use as a male contraceptive. It can be used to reduce heavy menstrual flow and in the treatment of endometriosis. The leaves are taken internally in the treatment of gastroenteritis. Externally, the leaves are used to treat thrush, scalds, bruises, and sores. The leaves are harvested as required during the growing season (Zargari, 2014; Mozaffarian, 2011; pfaf.org; Chikkulla *et al.*, 2018).

Hedera helix (Figure 2.55) is an evergreen climber that grows to a size of 15 m by 5 m at a medium rate. The species is hermaphrodite and pollinated by bees, flies, moths, and butterflies. It has a preference for light (sandy) and medium (loamy) soils. It can tolerate heavy clay and nutritionally poor soils. It can grow in acidic, neutral, basic (alkaline), and very alkaline soils, as well as in full shade (deep woodland), semi-shade (light woodland), or no shade. It prefers moist or wet soil and can tolerate drought. It can tolerate atmospheric pollution.

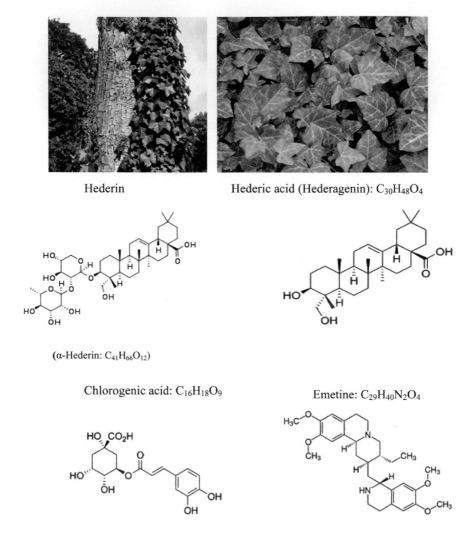

FIGURE 2.55 *Hedera helix* (whole plant and leaves) and its main components: Hederin, hederic acid, chlorogenic acid, emetine.

"Ivy" is a bitter aromatic herb with a nauseating taste. It is often used in folk herbal remedies, especially in the treatment of rheumatism and as an external application to skin eruptions, swollen tissue, painful joints, burns, and suppurating cuts. Recent research has shown that the leaves contain the compound "emetine," which is an amebicidal alkaloid, and also triterpene saponins, which are effective against liver flukes, molluscs, internal parasites, and fungal infections. The leaves are anti-bacterial, anti-rheumatic, antiseptic, anti-spasmodic, astringent, cathartic, diaphoretic, emetic, emmenagogue, stimulant, sudorific, vasoconstrictor, vasodilator, and vermifuge. The plant is used internally in the treatment of gout, rheumatic pain, whooping cough, bronchitis, and as a parasiticide. Some caution is advised if it is being used internally since the plant is mildly toxic. Excessive doses destroy red blood cells and cause irritability, diarrhea, and vomiting. This plant should only be used under the supervision of a qualified practitioner. An infusion of the twigs in oil is recommended for the treatment of sunburn. The leaves are used fresh and can also be dried reperfusion (Zargari, 2014; Mozaffarian, 2011; pfaf.org; Elias *et al.*, 1991).

Heracleum persicum (Figure 2.56) is a perennial that grows up to a size of 2 m. It is many stemmed, bristly haired, red-brown from the base, up to 50 mm thick, and with hollow joints with septa. It smells of aniseed. The flower is corolla regular, white, 15–30 mm wide; there are five petals, which are deeply notched. The sepals are stunted. There are five stamens. Pistil of two fused carpels, two styles. The bracts of primary umbels fall off early; the 10–18 bracteoles of secondary umbels do not fall off. The leaves are alternate and stalked, and the base is pod-like. The blade is longer than wide, with the lower side densely haired, glabrous on top, and pinnate. There are five to seven leaflets, which are big, shortly and broadly lobed, with blunt-toothed margins. The fruits are broadly obovate, 7–8 mm long, two-parted, with a slightly ridged schizocarp; the oilducts are only slightly club-shaped.

"Persian hogweed" (simply hogweed) has lots of medicinal properties and is used for the treatment of convulsions, inflammation, and fungal diseases. Recently, researchers have concluded that this plant has the potential for newer therapeutic applications in the future. Free radicals are considered as the major cause of various dangerous diseases of the heart and liver, Alzheimer's disease, cancer, and Parkinson's disease. This plant can be used as a free radical quencher since it has the ability to fight against it. As this plant has various useful compounds such as alkaloids, terpenoids, and triterpenes, it can treat convulsions quickly. It has anti-fungal, anti-microbial, and immunomodulatory activities as well. Many members of this genus, including many of the sub-species in this species, contain furanocoumarins, which have carcinogenic, mutagenic, and phototoxic properties. The juice from the plant is toxic and may cause skin reactions in exposed areas (phytotoxicity) when exposed to sunlight. This usually results in redness, itching, stinging, eczema, and the worst cases of severe burns (third-degree burns) as the toxins make the skin sensitive to UV rays. This sensitivity may persist for up to one year after exposure (Zargari, 2014; Mozaffarian, 2011; pfaf.org; Majidi & Sadati Lamardi, 2018).

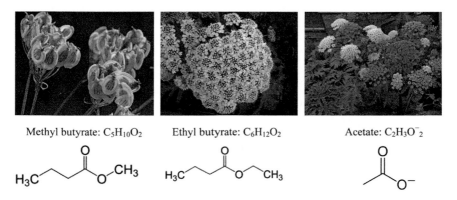

Methyl butyrate: $C_5H_{10}O_2$ Ethyl butyrate: $C_6H_{12}O_2$ Acetate: $C_2H_3O^-_2$

FIGURE 2.56 *Heracleum persicum* (seeds, flowers, and whole plant) and its main components: Methyl butyrate, ethyl butyrate, acetate.

Hypericum perforatum (Figure 2.57) is a perennial that grows to a size of 0.9 m by 0.6 m. The flowers are hermaphrodite and pollinated by bees and flies. The plant is self-fertile. It has a preference for light (sandy), medium (loamy), heavy (clay), and well-drained soil. It can grow in acidic, neutral, and basic (alkaline) soils, as well as in semi-shade (light woodland) or no shade. It prefers moist soil.

"St. John's wort" has a long history of herbal use. It fell out of favor in the nineteenth century, but recent research has brought it back to prominence as an extremely valuable remedy for nervous problems. In clinical trials, about 67% of patients with mild to moderate depression improved when taking this plant. The flowers and leaves are analgesic, antiseptic, anti-spasmodic, aromatic, astringent, cholagogue, digestive,

Hyperin: $C_{21}H_{19}O_{12}$ Hypericin: $C_{30}H_{16}O_8$

Pseudo-hypericin: $C_{30}H_{16}O_9$

FIGURE 2.57 *Hypericum perforatum* (flowers and whole plant) and main components: Hyperin, hypericin, pseudo-hypericin.

diuretic, expectorant, nervine, resolvent, sedative, stimulant, vermifuge, and vulnerary. The herb is used in treating a wide range of disorders, including pulmonary complaints, bladder problems, diarrhea, and nervous depression. It is also very effectual in treating overnight incontinence of urine in children. Externally, it is used in poultices to dispel herd tumors, caked breasts, bruising, etc. The flowering shoots are harvested in early summer and dried for later use. Use the plant with caution and do not prescribe it for patients with chronic depression. The plant was used to procure an abortion by some native North Americans, so it is best not used by pregnant women. See also the earlier notes on toxicity. A tea or tincture of the fresh flowers is a popular treatment for external ulcers, burns, wounds (especially those with severed nerve tissue), sores, bruises, cramps, etc. An infusion of the flowers in olive oil is applied externally to wounds, sores, ulcers, swellings, rheumatism, etc. It is also valued in the treatment of sunburn and as a cosmetic preparation to the skin. The plant contains many biologically active compounds including choline ($C_5H_{14}NO$), β-sitosterol ($C_{29}H_{50}O$), hypericin ($C_{30}H_{16}O_8$), and pseudo-hypericin ($C_{30}H_{16}O_9$). These last two compounds have been shown to have potent anti-retroviral activity without serious side effects and they are being researched in the treatment of AIDS. A homeopathic remedy is made from the fresh whole flowering plant. It is used in the treatment of injuries, bites, stings, etc., and is said to be the first remedy to consider when nerve-rich areas such as the spine, eyes, fingers, etc., are injured. Skin contact with the sap, or ingestion of the plant, can cause photosensitivity in some people. Common side effects are gastrointestinal disturbances, allergic reactions, and fatigue. If used with drugs classed as serotonin ($C_{10}H_{12}N_2O$) reuptake inhibitors, symptoms of serotonin syndrome may occur, including mental confusion, hallucinations, agitation, headache, coma, shivering, sweating, fever, hypertension, tachycardia, nausea, diarrhea, and tremors. St John's wort can reduce the effectiveness of prescription medicine, including the contraceptive pill, antidepressants, immune suppressants, HIV medications, warfarin, and digoxin (Zargari, 2014; Mozaffarian, 2011; pfaf.org; Huang *et al.*, 2011).

Illicium verum (Figure 2.58) is an evergreen tree that grows to a size of 5 m by 3 m. The species is hermaphrodite. It has a preference for light (sandy), medium (loamy), and well-drained soil. It can grow in acidic and neutral soils, as well as in semi-shade (light woodland) or no shade. It prefers moist soil.

The fruit of "star anise" is carminative, diuretic, odontalgic, stimulant, and stomachic. It is taken internally in the treatment of abdominal pain, digestive disturbances, and complaints such as lumbago. It is often included in remedies for digestive disturbances and cough mixtures, in part at least for its pleasant aniseed flavor. An effective remedy for various digestive upsets, including colic, it can be safely given to children. The fruit is also often chewed in small quantities after meals in order to promote digestion and to sweeten the breath. The fruit has an anti-bacterial affect similar to penicillin. The fruit is harvested unripe when used for chewing, the ripe fruits being used to extract essential oil and also dried for use in decoctions and powders. A homeopathic remedy is prepared from the seed (Zargari, 2014; Mozaffarian, 2011; pfaf.org; Paul Das & Sivagnanam, 2013).

Lawsonia inermis (Figure 2.59) or henna is a tall bushy tree or small tree, standing at 1.8 m to 7.6 m tall. It is glabrous and multi-branched, with spine-tipped branchlets. The leaves grow opposite each other on the stem: they are glabrous, sub-sessile, elliptical, and lanceolate (long and wider in the middle; average dimensions are

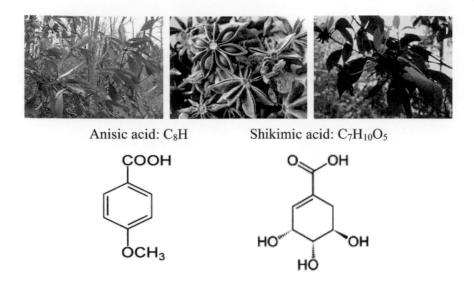

Anisic acid: C₈H   Shikimic acid: C₇H₁₀O₅

FIGURE 2.58 *Illicium verum* (leaves, fruits, and flowers) and its main components: Anisic acid, shikimic acid, and anethole (not shown).

FIGURE 2.59 *Lawsonia inermis* (flowers, seeds, and shrub).

1.5–5.0 × 0.5–2 cm), acuminate (tapering to a long point), and have depressed veins on the dorsal surface. Henna flowers have four sepals and a 2 mm calyxtube, with 3 mm spread lobes. Its petals are obvate, with white or red stamens found in pairs on the rim of the calyx tube. The ovary is four-celled, 5 mm long, and erect. Henna fruits are small, brownish capsules, 4–8 mm in diameter, with 32–49 seeds per fruit, and open irregularly into four splits.

"Henna" is useful for eczema and hand and foot sweat. It has an hepatoprotective and immunomodulatory effect, and anti-microbial, anthelminthic, anti-fungal, anti-trypanosomal, abortifacient, antioxidant, and anti-cancer activity of *Lawsonia inermis* were reported from all over the world by previous studies (Zargari, 2014; Mozaffarian, 2011; pfaf.org; Zumrutdal & Ozaslan, 2012).

Lens culinaris (Figure 2.60) is an annual that grows up to a size of 0.5 m. It is not frost tender. The flowers are hermaphrodite and pollinated by cleistogamous. The plant is self-fertile. It can fix nitrogen. It has a preference for light (sandy), medium (loamy), heavy (clay), and well-drained soil, and can grow in nutritionally poor soil. It can grow in acid, neutral, and basic (alkaline) soils. It cannot grow in the shade. It prefers dry or moist soil.

Starch: $(C_6H_{10}O_5)_n - (H_2O)$

(Amylose)

FIGURE 2.60 *Lens culinaris* (pod, seeds, and whole plant) and its main components: Starch, and vitamin B_9 (not shown).

The seeds of the "lentil" are mucilaginous and laxative. They are considered to be useful in the treatment of constipation and other intestinal affections. Made into a paste, they are a useful cleansing application in foul and indolent ulcers (Zargari, 2014; Mozaffarian, 2011; pfaf.org; Vohra & Gupta, 2012).

Lepidium latifolium (Figure 2.61) is a perennial that grows to a size of 1.2 m by 1 m. The species is hermaphrodite and pollinated by insects. It has a preference for light (sandy), medium (loamy), and heavy (clay) soils. It can grow in acidic, neutral, basic (alkaline), and saline soils, as well as in semi-shade (light woodland) or no shade. It prefers moist soil. The plant can tolerate maritime exposure.

Lepidine (4-methylquinoline): $C_{10}H_9N$

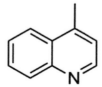

FIGURE 2.61 *Lepidium latifolium* (whole plant, leaves, and flowers) and its main component: Lepidine.

"Tall whitetop" is anti-scorbutic, depurative, and stomachic. An infusion of the plant is used in the treatment of liver and kidney diseases. It also increases cardiac amplitude, decreases frequency, and regulates the rhythm. It is also used as a resolvent in the treatment of skin diseases (Zargari, 2014; Mozaffarian, 2011; pfaf.org; Kaur *et al.*, 2013).

Lepidium sativum (Figure 2.62) is an annual that grows up to a size of 0.5 m. The species is hermaphrodite and pollinated by insects. It has a preference for light (sandy), medium (loamy), and heavy (clay) soils. It can grow in acidic, neutral, and basic (alkaline) soils, as well as in semi-shade (light woodland) or no shade. It prefers moist soil.

The leaves of "garden cress" are anti-scorbutic, diuretic, and stimulant. The plant is administered in cases of asthma, cough with expectoration, and bleeding piles. The root is used in the treatment of secondary syphilis and tenesmus. The seeds are galactogogue. They have been boiled with milk and used to procure an abortion, have been applied as a poultice to aches and pains, and have also been used as an aperient. Fresh foliage is 37% vitamin C ($C_6H_8O_6$) (Zargari, 2014; Mozaffarian, 2011; pfaf.org; Naval & Pandya, 2011).

linum usitatissimum (Figure 2.63) is an annual that grows to a size of 0.7 m by 0.2 m. The species is hermaphrodite and pollinated by insects. The plant is self-fertile. It has a preference for light (sandy), medium (loamy), and well-drained soil. It can grow in acidic, neutral, and basic (alkaline) soils. It cannot grow in the shade. It prefers moist soil. The plant can tolerate strong winds but not maritime exposure.

"Linseed" has a long history of medicinal use, its main effects being as a laxative and expectorant that soothes irritated tissues, controls coughing, and relieves pain. The seed, or the oil from the seed, is normally used. The seed is analgesic, demulcent, emollient, laxative, pectoral, and resolvent. The crushed seed makes a very useful poultice in the treatment of ulceration, abscesses, and deep-seated inflammations. An infusion of the seed contains a good deal of mucilage and is a valuable domestic remedy for coughs, colds, and inflammation of the urinary organs. If the seed is bruised and then eaten straightaway, it will swell considerably in the digestive tract and stimulate peristalsis and so is used in the treatment of chronic constipation. The oil in the seed contains 4% L-glutamic acid ($C_5H_9NO_4$), which is used to treat mental deficiencies in adults. It also has soothing and lubricating properties, and is used in medicines to soothe tonsillitis, sore throats, coughs, colds, constipation, gravel, and stones. When mixed with an equal quantity of lime water, it is used to treat burns and

FIGURE 2.62   *Lepidium sativum* (seeds, leaves, and flowers).

Linamarin: $C_{10}H_{17}NO_6$

FIGURE 2.63 *Linum usitatissimum* (whole plant and flower) and a main components: Linamarin,.

scalds. The bark and the leaves are used in the treatment of gonorrhea. The flowers are cardiotonic and nervine (Zargari, 2014; Mozaffarian, 2011; pfaf.org; Chantreau *et al.*, 2015).

Malus orientalis (Figure 2.64) is a deciduous tree that grows up to a size of 7 m at a medium rate. The flowers are hermaphrodite and pollinated by insects. It is noted for attracting wildlife. It has a preference for light (sandy), medium (loamy), and well-drained soil, and can tolerate heavy clay soil. It can grow in acidic, neutral, and basic (alkaline) soils, as well as in semi-shade (light woodland) or no shade. It prefers moist soil.

The bark, and especially the root bark of the "apple" is anthelmintic, refrigerant, and soporific. An infusion is used in the treatment of intermittent, remittent, and bilious fevers. The fruit is said to dispel gas, dissolve mucous, treat flux, and be a tonic for anemia, bilious disorders, and colic. The leaves contain up to 2.4% of an anti-bacterial substance called "phloretin," which inhibits the growth of a number of gram-positive and gram-negative bacteria in as low a concentration as 30 ppm. The plant is used in Bach flower remedies—the keywords for prescribing it are the "cleansing remedy," "despondency," and "despair." All members of this genus contain the toxin HCN in their seeds and possibly also in their leaves, but not in their fruits. HCN is the substance that gives almonds their characteristic taste, but it should only be consumed in very small quantities. Apple seeds do not normally contain very high quantities of HCN but, even so, should not be consumed in very large quantities. In small quantities, HCN has been shown to stimulate respiration and improve digestion;

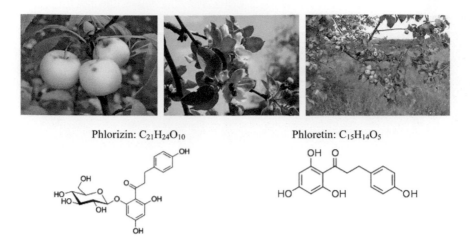

Phlorizin: $C_{21}H_{24}O_{10}$ Phloretin: $C_{15}H_{14}O_5$

FIGURE 2.64 *Malus orientalis* (fruits, blossoms, and tree) and its main components: Phlorizin, phloretin.

it is also claimed to be of benefit in the treatment of cancer. In excess, however, it can cause respiratory failure and even death (Zargari, 2014; Mozaffarian, 2011; pfaf.org; Monika *et al.*, 2013).

Mangifera indica (Figure 2.65), otherwise known as mango, is an evergreen tree that grows to a size of 25 m by 25 m at a medium rate. The flowers are pollinated by bees, bats, flies, and ants. The plant is self-fertile. It has a preference for light (sandy), medium (loamy), heavy (clay), and well-drained soil. It can grow in acidic (including very acidic), neutral, and saline soils. It cannot grow in the shade. It prefers moist soil and can tolerate drought. It is a popular fruit tree and the national tree of Bangladesh. It is large and evergreen growing up to 45 m in height and 120 cm in bole diameter. The canopy is umbrella-shaped and spreading. It has a taproot system that can be up to 5 m deep. It is commonly grown in East Asia. Mango fruit is the national fruit of India, Pakistan, and the Philippines. One of the most popular fruits worldwide, it can be eaten raw, processed into juice, jams, candies, etc., or dried and ground into powder. The seeds are sources of starch and edible fat. Young leaves are cooked as a vegetable. The bark and leaves yield yellowish-brown dye used for silk. The flowers are used to repel mosquitos. The wood is used for construction, furniture, carpentry, flooring, boxes, and crates, etc. It is moderately heavy, moderately hard, not durable under exposed conditions, and susceptible to fungi, borers, and termites.

"Mango" has medicinal uses as well. Generally, it is anti-diuretic, anti-diarrheal, and anti-emetic. Plant parts are used in the treatment of various conditions such as high blood pressure, angina, asthma, coughs, diabetes, dental problems, skin problems, colds, diarrhea, bleeding piles, dysentery, scorpion stings, hemorrhage, stomach pain, etc. An infusion is drunk to reduce blood pressure and as a treatment for conditions such as angina, asthma, coughs, and diabetes. Externally, the leaves are used in a convalescent bath. A mouthwash made from the leaves is effective in hardening the gums and helping to treat dental problems. The leaves are used to treat skin irritations. The charred and pulverized leaves are used to make a plaster for removing warts and also act as a styptic. The seed is astringent, anti-diarrheal; it is anthelmintic

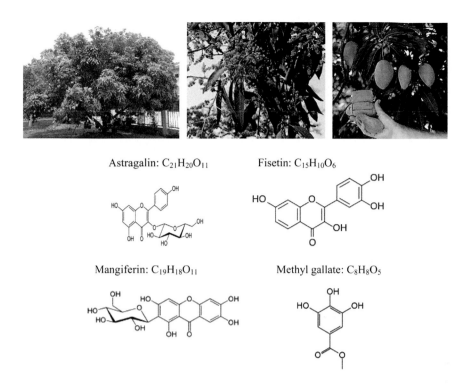

Astragalin: $C_{21}H_{20}O_{11}$ Fisetin: $C_{15}H_{10}O_6$

Mangiferin: $C_{19}H_{18}O_{11}$ Methyl gallate: $C_8H_8O_5$

FIGURE 2.65 *Mangifera indica* (tree, leaves, and fruits) and its main components: Astragalin, fisetin, mangiferin, methyl gallate.

when roasted. It is used to treat stubborn colds and coughs, obstinate diarrhea, and bleeding piles. The pulverized seed is made into a sweetened tea and drunk, or taken as powders, for treating dysentery. The seeds are ground up and used to treat scorpion stings. The bark is astringent, homeostatic, and anti-rheumatic, and is used in the treatment of hemorrhage, diarrhea, and throat problems. When incised, the bark yields an oleoresin which is stimulant, sudorific, and anti-syphilitic. The stem is astringent and is used to treat diarrhea and to remedy stomachache. The roots are diuretic. The flowers are aphrodisiac. The fruit is anti-scorbutic and anti-dysenteric (Zargari, 2014; Mozaffarian, 2011; pfaf.org; Vieccelli *et al.*, 2016).

Medicago sativa (Figure 2.66) is a perennial that grows up to a size of 1 m at a medium rate. The flowers are hermaphrodite and pollinated by bees. The plant is self-fertile. It can fix nitrogen. It is noted for attracting wildlife. It has a preference for light (sandy), medium (loamy), heavy (clay), and well-drained soil, and can tolerate nutritionally poor soil. It can grow in acidic, neutral, and basic (alkaline) soils. It cannot grow in the shade. It prefers dry or moist soil and can tolerate drought.

"Alfalfa" leaves, either fresh or dried, have traditionally been used as a nutritive tonic to stimulate the appetite and promote weight gain. The plant has an estrogenic action and could prove useful in treating problems related to menstruation and the menopause. Some caution is advised in the use of this plant, however. It should not be prescribed to people with auto-immune diseases such as rheumatoid arthritis. See also the earlier notes on toxicity. The plant is anti-scorbutic, aperient, diuretic,

Asparagine: $C_4H_8N_2O_3$

Chlorophyll: $C_{55}H_{72}O_5N_4Mg$

(Chlorophyll *c1*)

Tricin: $C_{17}H_{14}O_7$ Canavanine: $C_5N_4H_{12}O_3$

FIGURE 2.66 *Medicago sativa* (whole plant and flower) and its main components: Asparagine, chlorophyll, tricin, canavanine.

oxytocic, hemostatic, nutritive, stimulant, and tonic. The expressed juice is emetic and also anodyne in the treatment of gravel. The plant is taken internally for debility in convalescence or anemia, hemorrhage, menopausal complaints, pre-menstrual tension, fibroids, etc. A poultice of the heated leaves can be applied to the ear in the treatment of earache. The leaves can be used fresh or dried. The leaves are rich in vitamin K which is used medicinally to encourage the clotting of blood. This is valuable in the treatment of jaundice. The plant is grown commercially as a source of chlorophyll and carotene, both of which have proven health benefits. The leaves also contain the antioxidant tricin ($C_{17}H_{14}O_7$). The root is febrifuge and is also prescribed in cases of highly colored urine. Extracts of the plant are anti-bacterial. It is used for asthma, diabetes, gastrointestinal disorders (anti-ulcer). The plant contains saponin-like substances. Eating large quantities of the leaves may cause the breakdown of red blood cells. However, although they are potentially harmful, saponins are poorly absorbed by the human body and so most pass through without harm. Saponins are quite bitter and can be found in many common foods such as some beans; thorough cooking, and perhaps changing the cooking water once, will normally remove most of them from the food. Saponins are much more toxic to some creatures, such as fish, and hunting tribes have traditionally put large quantities of them in streams, lakes, etc., in order to stupefy or kill the fish. Alfalfa sprouts and especially the seeds contain canavanine ($C_5N_4H_{12}O_3$). Recent reports suggest that ingestion of this substance can cause the recurrence of systemic lupus erythematosus in patients where the disease had become dormant. It is advised that children, the elderly, and people with compromised immune systems should avoid eating alfalfa sprouts due to bacterial contamination. They should also be avoided during pregnancy and lactation, as well for people with hormone-sensitive cancer. Also, they possibly antagonize the anti-coagulant effect of warfarin and interfere with the immunosuppressant effect of corticosteroids (Zargari, 2014; Mozaffarian, 2011; pfaf.org; Bora & Sharma, 2011).

Melilotus officinalis (Figure 2.67) is an annual/biennial that grows to a size of 1.2 m by 0.7 m. It is not frost tender. The flowers are hermaphrodite and pollinated by bees. It can fix nitrogen. It has a preference for light (sandy), medium (loamy), and well-drained soil, and can also tolerate heavy clay soil. It can grow in neutral, basic (alkaline), and saline soils. It cannot grow in the shade. It prefers dry or moist soil and can tolerate drought.

"Melilot," used either externally or internally, can help treat varicose veins and hemorrhoids though it requires a long-term treatment for the effect to be realised. Use of the plant also helps to reduce the risk of phlebitis and thrombosis. Melilot contains coumarin ($C_9H_6O_2$) and, as the plant dries or spoils, this become converted to dicoumarol ($C_{19}H_{12}O_6$), a powerful anti-coagulant. Thus, the plant should be used with some caution: it should not be prescribed to patients with a history of poor blood clotting or who are taking warfarin medication. See also the earlier notes on toxicity. The flowering plant is anti-spasmodic, aromatic, carminative, diuretic, emollient, mildly expectorant, mildly sedative, and vulnerary. An infusion is used in the treatment of sleeplessness, nervous tension, neuralgia, palpitations, varicose veins, painful congestive menstruation, in the prevention of thrombosis, flatulence, and intestinal disorders. Externally, it is used to treat eye inflammations, rheumatic pains, swollen joints, severe bruising, boils, and erysipelas, where a decoction is added to the bathwater. The flowering plant is harvested in the summer and can be dried for later

Coumarin: $C_9H_6O_2$　　　　　　　　Melilotic acid: $C_9H_{10}O_3$

Dicoumarol: $C_{19}H_{12}O_6$

FIGURE 2.67 *Melilotus officinalis* (flowers and whole plant) and its main components: Coumarin, melilotic acid, dicoumarol.

use. A distilled water obtained from the flowering tops is an effective treatment for conjunctivitis. The dried leaves can be toxic, though the fresh leaves are quite safe to use, possibly due to the presence of coumarin ($C_9H_6O_2$), the substance that gives some dried plants the smell of freshly cut hay. If taken internally it can prevent the blood clotting (Zargari, 2014; Mozaffarian, 2011; pfaf.org; Sheikh *et al.*, 2016).

Mespilus germanica (Figure 2.68) is a deciduous tree that grows up to a size of 6 m at a medium rate. The flowers are hermaphrodite and pollinated by bees. The plant is self-fertile. It has a preference for light (sandy), medium (loamy), heavy (clay), and well-drained soil. It can grow in acidic, neutral, and basic (alkaline) soils, as well as in semi-shade (light woodland) or no shade. It prefers moist soil. The plant can tolerate strong winds but not maritime exposure.

The pulp of the fruit of "common medlar" is laxative. The leaves are astringent. The seed is lithontripic. It is ground up for use, but caution should be employed since the seeds contain the toxin HCN. The bark has been used as a substitute for quinine ($C_{20}H_{24}N_2O_2$), but with uncertain results. As mentioned, the seeds contain the toxic

Quinine: $C_{20}H_{24}N_2O_2$

FIGURE 2.68 *Mespilus germanica* (fruits and flower) and a main component: Quinine.

HCN (the substance that gives almonds their flavor) and should not be eaten in quantity (Zargari, 2014; Mozaffarian, 2011; pfaf.org; Ercisli *et al.*, 2012).

Myrtus communis (Figure 2.69) is an evergreen bushy tree that grows to a size of 4.5 m by 3 m at a medium rate. The flowers are hermaphrodite and pollinated by bees. The plant is self-fertile. It has a preference for light (sandy), medium (loamy), heavy (clay), and well-drained soil. It can grow in acidic, neutral, and basic (alkaline) soils. It cannot grow in the shade. It prefers dry or moist soil. The plant can tolerate maritime exposure.

The leaves of "myrtle" are aromatic, balsamic, hemostatic, and tonic. Recent research has revealed a substance in the plant that has an antibiotic action. The active ingredients are rapidly absorbed and give a violet-like scent to the urine within 15 minutes. The plant is taken internally in the treatment of urinary infections, digestive problems, vaginal discharge, bronchial congestion, sinusitis, and dry coughs. In India it is useful in the treatment of cerebral affections, especially epilepsy. Externally, it is used in the treatment of acne (the essential oil is normally used here), wounds, gum infections, and hemorrhoids. The leaves are picked as required and used fresh or dried. An essential oil obtained from the plant is antiseptic. It is used as a remedy for gingivitis. The oil is used as a local application in the treatment of rheumatism. The fruit is carminative. It is used in the treatment of dysentery, diarrhea,

Linolein: $C_{57}H_{98}O_6$

FIGURE 2.69 *Myrtus communis* (flowers, whole plant, and fruits) and its main component: Linolein.

hemorrhoids, internal ulceration, and rheumatism (Zargari, 2014; Mozaffarian, 2011; pfaf.org; Hennia *et al.*, 2018).

Nasturtium officinale (Figure 2.70) is a perennial that grows to a size of 0.5 m by 1 m at a fast rate. The flowers are hermaphrodite and pollinated by bees and flies. The plant is self-fertile. It is noted for attracting wildlife. It has a preference for light (sandy), medium (loamy), and heavy (clay) soils. It can grow in acidic, neutral, and basic (alkaline) soils. It cannot grow in the shade. It prefers wet soil and can grow in water.

Gluconasturtiin: $C_{15}H_{21}NO_9S_2$ Phenyl ethyl (2-phenylethanol): $C_8H_{10}O$

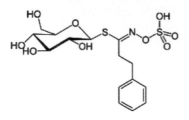

FIGURE 2.70 *Nasturtium officinale* (flowers and whole plant) and its main components: Gluconasturtiin, phenyl ethyl, and vitamin C (not shown).

"Watercress" is very rich in vitamins and minerals, and has long been valued as a food and medicinal plant. Considered a cleansing herb, its high content of vitamin C ($C_6H_8O_6$) makes it a remedy that is particularly valuable for chronic illnesses. The leaves are anti-scorbutic, depurative, diuretic, expectorant, purgative, hypoglycemic, odontalgic, stimulant, and stomachic. The plant has been used as a specific in the treatment of TB. The freshly pressed juice has been used internally and externally in the treatment of chest and kidney complaints, chronic irritations, and inflammations of the skin, etc. Applied externally, it has a long-standing reputation as an effective hair tonic, helping to promote the growth of thick hair. A poultice of the leaves is said to be an effective treatment for healing glandular tumors or lymphatic swellings. Some caution is advised as excessive use of the plant can lead to stomach upsets. The leaves can be harvested almost throughout the year and are used fresh. Whilst the plant is very wholesome and nutritious, some care should be taken if harvesting it from the wild. Any plants growing in water that drains from fields where animals, particularly sheep, graze should not be used raw. This is due to the risk of it being infested with the liver fluke parasite. Cooking the leaves, however, will destroy any parasites and render the plant perfectly safe to eat. It may inhibit the metabolism of paracetamol (Zargari, 2014; Mozaffarian, 2011; pfaf.org; Voutsina *et al.*, 2016).

Nigella sativa (Figure 2.71) is an annual that grows to a size of 0.4 m by 0.2 m. The flowers are hermaphrodite and pollinated by bees. It has a preference for light (sandy), medium (loamy), and heavy (clay) soils, as well as well-drained soil. It can grow in acidic, neutral, and basic (alkaline) soils. It cannot grow in the shade. It prefers dry or moist soil.

Nigellone (Polythymoquinone): $C_{20}H_{24}O_4$

FIGURE 2.71 *Nigella sativa* (flowers and seeds) and its main component: Nigellone.

Like many aromatic culinary herbs, the seeds of "black cumin" are beneficial for the digestive system, soothing stomach pains and spasms, as well as easing wind, bloating, and colic. The ripe seed is anthelmintic, carminative, diaphoretic, digestive, diuretic, emmenagogue, galactogogue, laxative, and stimulant. An infusion is used in the treatment of digestive and menstrual disorders, insufficient lactation, and bronchial complaints. The seeds are much used in India to increase the flow of milk in nursing mothers and they can also be used to treat intestinal worms, especially in children. Externally, the seed is ground into a powder, mixed with sesame oil and used to treat abscesses, hemorrhoids, and orchitis. The powdered seed has been used to remove lice from the hair (Zargari, 2014; Mozaffarian, 2011; pfaf.org; Hayat, 2013).

Nymphaea alba (Figure 2.72) is a perennial. The species is hermaphrodite and pollinated by flies and beetles. The plant is self-fertile. It has a preference for light (sandy), medium (loamy), and heavy (clay) soils. It can grow in acidic, neutral, and basic (alkaline) soils. It cannot grow in the shade. It can grow in water.

The rhizome of the "white water-lily" is anodyne, anti-scrofulatic, astringent, cardiotonic, demulcent, and sedative. A decoction of the root is used in the treatment of dysentery or diarrhea caused by irritable bowel syndrome. It has also been used to treat bronchial catarrh and kidney pain and can be taken as a gargle for sore throats. Externally, it can be used to make a douche to treat vaginal soreness or discharges. In combination with slippery elm (*Ulmus rubra*) or flax (*Linum usitatissimum*), it is used as a poultice to treat boils and abscesses. The rhizome is harvested in the autumn and can be dried for later use. The flowers are anaphrodisiac and sedative and generally have a calming and sedative effect upon the nervous system, reputedly reducing the

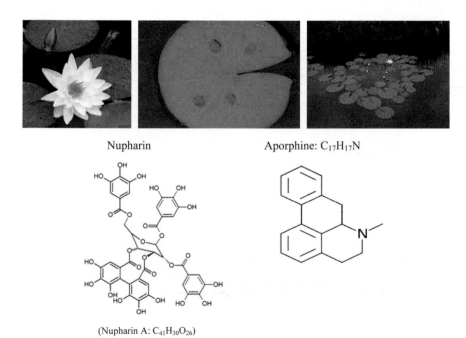

Nupharin

Aporphine: $C_{17}H_{17}N$

(Nupharin A: $C_{41}H_{30}O_{26}$)

FIGURE 2.72 *Nymphaea alba* (flower, leaf, and whole plant) and its main components: Nupharin, aporphine.

sex drive and making them useful in the treatment of insomnia, anxiety, and similar disorders. A treatment of uterine cancer by a decoction and uterine injection has been recorded. According to one report the plant is not used in modern herbal practice, though it has been quoted as a remedy for dysentery (Zargari, 2014; Mozaffarian, 2011; pfaf.org; Omar Bakr *et al.*, 2017).

Onobrychis viciifolia (Figure 2.73) is a perennial that grows up to a size of 1 m. The species is hermaphrodite and pollinated by bees. The plant is not self-fertile. It can fix nitrogen. It is noted for attracting wildlife. It has a preference for light (sandy), medium (loamy) soils, and well-drained soil, and can tolerate nutritionally poor soil. It can grow in neutral and basic (alkaline) soils, as well as in very alkaline soils. It cannot grow in the shade. It prefers moist soil.

The leaves of "sainfoin" contain high levels of condensed tannins. Sainfoin is used as sweaty and appetizer but it is not common (Zargari, 2014; Mozaffarian, 2011; pfaf.org; Jones & Mangan, 1977).

Oxalis corniculata (Figure 2.74) is an annual/perennial that grows to a size of 0.1 m by 0.3 m. The species is hermaphrodite and pollinated by insects. The plant is self-fertile. It has a preference for light (sandy), medium (loamy), heavy (clay), and well-drained soil. It can grow in acidic, neutral, and basic (alkaline) soils. It cannot grow in the shade. It prefers dry or moist soil.

"Creeping woodsorrel" (yellow sorrel) is anthelmintic, antiphlogistic, astringent, depurative, diuretic, emmenagogue, febrifuge, lithontripic, stomachic, and styptic. It is used in the treatment of influenza, fever, urinary tract infections, enteritis, diarrhea, traumatic injuries, sprains, and poisonous snake bites. The juice of the plant, mixed with butter, is applied to muscular swellings, boils, and

FIGURE 2.73 *Onobrychis viciifolia* (flower and whole plant).

FIGURE 2.74 *Oxalis corniculata* (flower, leaves, and whole plant).

pimples. An infusion can be used as a wash to rid children of hookworms. The plant is a good source of vitamin C and is used as an anti-scorbutic in the treatment of scurvy. The leaves are used as an antidote to poisoning by the seeds of *Datura spp.*, arsenic, and mercury. The leaf juice is applied to insect bites, burns, and skin eruptions. It has an anti-bacterial activity (Zargari, 2014; Mozaffarian, 2011; pfaf.org; Raghavendra *et al.*, 2006).

Paeonia corralina (P. mascula) (Figure 2.75) is a perennial that grows to a size of 1 m by 1 m. The species is hermaphrodite and pollinated by insects. The plant is self-fertile. It has a preference for light (sandy) and medium (loamy), and can tolerate heavy clay soil. It can grow in acidic, neutral, and basic (alkaline) soils, as well as in semi-shade (light woodland) or no shade. It prefers dry or moist soil and can tolerate drought.

The root of "wild peony" is anti-spasmodic and tonic. A tea made from the dried crushed petals of various peony species has been used as a cough remedy, and as a treatment for hemorrhoids and varicose veins (Zargari, 2014; Mozaffarian, 2011; pfaf.org; Mozaffarian, 1996; Yu Ding *et al.*, 2000).

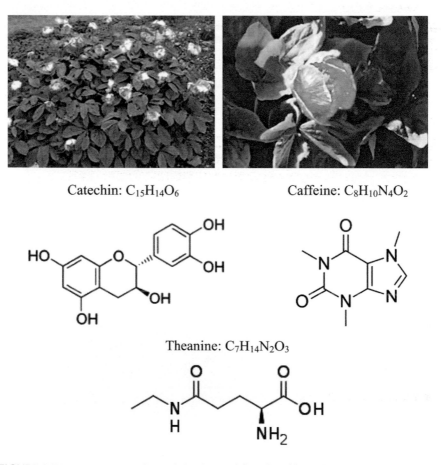

Catechin: $C_{15}H_{14}O_6$ Caffeine: $C_8H_{10}N_4O_2$

Theanine: $C_7H_{14}N_2O_3$

FIGURE 2.75   *Paeonia corralina* (whole plant and flower) and its main components: Catechin, caffeine, theanine.

Papaver rhoeas (Figure 2.76) is an annual that grows to a size of 0.6 m by 0.2 m at a fast rate. The flowers are hermaphrodite and pollinated by bees, flies, and beetles. The plant is self-fertile. It is noted for attracting wildlife. It has a preference for light (sandy), medium (loamy), and heavy (clay) soils, as well as well-drained soil. It can grow in acidic, neutral, and basic (alkaline) soils. It cannot grow in the shade. It prefers moist soil.

The flowers of the "corn poppy" have a long history of medicinal usage, especially for ailments in the elderly and children. Chiefly employed as a mild pain reliever and as a treatment for irritable coughs, it also helps to reduce nervous over-activity. Unlike the related opium poppy (*P. somniferum*), it is non-addictive. However, the plant does contain alkaloids, which are still being investigated, and so should only be used under the supervision of a qualified herbalist. The flowers and petals are anodyne, emollient, emmenagogue, expectorant, hypnotic, slightly narcotic, and sedative. An infusion is taken internally in the treatment of bronchial complaints and coughs, insomnia, poor digestion, nervous digestive disorders, and minor painful conditions. The flowers are also used in the treatment of jaundice. The petals are harvested as the flowers open and are dried for later use; they should be collected on a dry day and can be dried or made into a syrup. The latex in the seedpods is narcotic and slightly sedative. It can

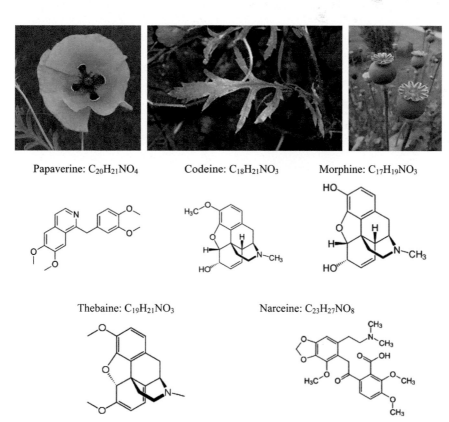

Papaverine: $C_{20}H_{21}NO_4$ Codeine: $C_{18}H_{21}NO_3$ Morphine: $C_{17}H_{19}NO_3$

Thebaine: $C_{19}H_{21}NO_3$ Narceine: $C_{23}H_{27}NO_8$

FIGURE 2.76 *Papaver rhoeas* (flower, leaf, and seedpods) and its main components: Papaverine, codeine, morphine, thebaine, narceine.

be used in very small quantities, and under expert supervision, as a sleep-inducing drug. The leaves and seeds are tonic. They are useful in the treatment of low fevers. The plant has anti-cancer properties and is toxic to mammals, though the toxicity is low. The seed is not toxic (Zargari, 2014; Mozaffarian, 2011; pfaf.org; Hasplova *et al.*, 2011).

Papaver somniferum (Figure 2.77), commonly known as the "opium poppy," is an annual herb that grows to a size of 0.6 m by 0.2 m. The species is hermaphrodite and pollinated by bees. The plant is self-fertile. It has a preference for light (sandy), medium (loamy), heavy (clay), and well-drained soil. It can grow in acidic, neutral, and basic (alkaline) soils. It cannot grow in the shade. It prefers moist soil. The stem and leaves bear a sparse distribution of coarse hairs. The flowers normally have four white, mauve, or red petals. The fruit is a hairless, rounded capsule topped with 12–18 radiating stigmatic rays, or fluted cap. All parts of the plant exude white latex when wounded. The opium poppy, as its name indicates, is the principal source of opium, the dried latex produced by the seedpods. It is extracted by making shallow incisions in the capsules as soon as the petals have fallen.

The latex exudes from the capsules and dries in contact with the air, and is then scraped off. This latex is anodyne, anti-tussive, astringent, diaphoretic, emmenagogue, hypnotic, narcotic, and sedative. As well as its pain-relieving properties, the latex has also been used as an anti-spasmodic and expectorant in treating certain kinds of coughs, whilst its astringent properties make it useful in the treatment of dysentery, etc. The opiate drugs are extracted from opium. Opium contains a class of naturally occurring alkaloids known as opiates, which include morphine ($C_{17}H_{19}NO_3$), thebaine

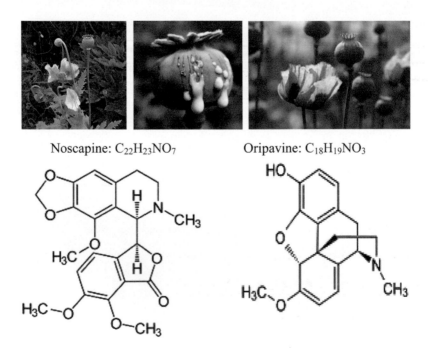

Noscapine: $C_{22}H_{23}NO_7$ Oripavine: $C_{18}H_{19}NO_3$

FIGURE 2.77 *Papaver somniferum* (whole plant, seedpod with latex flowing, and flower) and its main components: Noscapine, oripavine.

($C_{19}H_{21}NO_3$), codeine ($C_{18}H_{21}NO_3$), papaverine ($C_{20}H_{21}NO_4$), noscapine ($C_{22}H_{23}NO_7$), and oripavine ($C_{18}H_{19}NO_3$). The seed, raw or cooked, has been used as a flavoring in cakes and bread. In addition, the crushed and sweetened seeds were used as a filling in pastries. But all uses of opium are forbidden in Iran. Although the trading, shipping, and use of its products are prohibited in Iran and come with serious fines, unfortunately, some people are addicted to its products which are used as narcotics. Among its products, opium (Figure 2.78), heroin (Figure 2.79), and the juice of opium (Figure 2.80) are more favored. Opium is taken via shisha, tube, injection, bafoor (a special device for smoking opium which is used with charcoal) (Figure 2.81), and eating. Heroin is taken via tube, injection, and inhalation. In addition, the juice of opium is taken via tube, injection, and eating. The majority of poppy products are smuggled from Afghanistan and Pakistan to Iran, but in some places, people plant poppy plants illegally in very small areas (Zargari, 2014; Mozaffarian, 2011; pfaf.org; Saberi Zafarghandi *et al.*, 2015; Aliverdinia & Pridemore, 2008).

FIGURE 2.78   Opium.

FIGURE 2.79   Heroin.

FIGURE 2.80   Juice of opium.

FIGURE 2.81   Bafoor.

Peganum harmala (Figure 2.82) is a perennial that grows to a size of 0.6 m by 0.5 m. The flowers are hermaphrodite. It has a preference for light (sandy) medium (loamy), and well-drained soil. It can grow in acidic, neutral, basic (alkaline), and saline soils. It cannot grow in the shade. It prefers dry or moist soil.

The fruit and seed of "peganum" are digestive, diuretic, hallucinogenic, narcotic, and uterine stimulant. They are taken internally in the treatment of stomach complaints, urinary and sexual disorders, epilepsy, menstrual problems, and mental and nervous illnesses. The seed has also been used as an anthelmintic in order to rid the body of tapeworms. This remedy should be used with caution and preferably under the guidance of a qualified practitioner as excessive doses cause vomiting and hallucinations. The seeds contain the substance harmine ($C_{13}H_{12}N_2O$) which is being used in the research into mental disease, encephalitis, and inflammation of the brain.

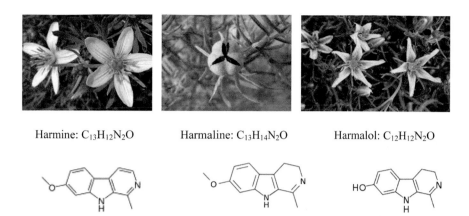

Harmine: $C_{13}H_{12}N_2O$ Harmaline: $C_{13}H_{14}N_2O$ Harmalol: $C_{12}H_{12}N_2O$

FIGURE 2.82 *Peganum harmala* (flowers, seedpod, and whole plant) and its main components: Harmine, harmaline, harmalol.

Small quantities stimulate the brain and are said to be therapeutic, but in excess, harmine ($C_{13}H_{12}N_2O$) depresses the central nervous system. A crude preparation of the seed is more effective than an extract because of the presence of related indoles. Consumption of the seed in quantity induces a sense of euphoria and releases inhibitions. It has been used in the past as a truth drug. The oil obtained from the seed is said to be an aphrodisiac. The oil is also said to have galactogogue, ophthalmic, soporific, and vermifuge properties. The seed is used externally in the treatment of hemorrhoids and baldness. The whole plant is said to be abortifacient, aphrodisiac, emmenagogue, and galactogogue. A decoction of the leaves is used in the treatment of rheumatism. The root has been used as a parasiticide in order to kill body lice. It is also used internally in the treatment of rheumatism and nervous conditions. Use with caution. Although the seed is used medicinally and as a condiment, it does contain hallucinogenic and narcotic alkaloids. When taken in excess it causes hallucinations and vomiting (Zargari, 2014; Mozaffarian, 2011; pfaf.org; Ayoob *et al.*, 2017).

Persica vulgaris (prunus persica) (Figure 2.83) is a fast-growing deciduous tree reaching up to 6 m. The flowers are hermaphrodite and pollinated by bees. The plant is self-fertile. It has a preference for light (sandy), medium (loamy), heavy (clay), and

FIGURE 2.83 *Persica vulgaris* (flowers and fruits).

well-drained soil. It can grow in acidic, neutral, and basic (alkaline) soils. It cannot grow in the shade. It prefers moist soil.

The leaves of the "nectarine" are astringent, demulcent, diuretic, expectorant, febrifuge, laxative, parasiticide, and mildly sedative. They are used internally in the treatment of gastritis, whooping cough, coughs, and bronchitis. They also help to relieve vomiting and morning sickness during pregnancy, though the dose must be carefully monitored because of their diuretic action. The dried and powdered leaves have sometimes been used to help heal sores and wounds. The flowers are diuretic, sedative, and vermifuge. They are used internally in the treatment of constipation and edema. A gum from the stems is alterative, astringent, demulcent, and sedative. The seed is anti-asthmatic, anti-tussive, emollient, hemolytic, laxative, and sedative. It is used internally in the treatment of constipation in the elderly, coughs, asthma, and menstrual disorders. The bark is demulcent, diuretic, expectorant, and sedative. It is used internally in the treatment of gastritis, whooping cough, coughs, and bronchitis. The root bark is used in the treatment of dropsy and jaundice. The bark is harvested from young trees in the spring and is dried for later use. The seed contains "laetrile," a substance that has also been called vitamin B_{17} ($C_{20}H_{27}NO_{11}$). This has been claimed to have a positive effect in the treatment of cancer, but at present there does not seem to be much evidence to support this. The pure substance is almost harmless, but on hydrolysis it yields hydrocyanic acid, a very rapidly acting poison—it should thus be treated with caution. In small amounts, this exceedingly poisonous compound stimulates respiration, improves digestion, and gives a sense of well-being. The seed can contain high levels of hydrocyanic acid (HCN), a poison that gives almonds their characteristic flavor. This toxin is readily detected by its bitter taste. Usually present in too small a quantity to do any harm, any very bitter seed or fruit should not be eaten. In small quantities, HCN has been shown to stimulate respiration and improve digestion; it is also claimed to be of benefit in the treatment of cancer. In excess, however, it can cause respiratory failure and even death (Zargari, 2014; Mozaffarian, 2011; pfaf.org; Mozaffarian, 1996; Hussain *et al.*, 2013).

Petroselinum crispum (Figure 2.84) is a biennial that grows to a size of 0.6 m by 0.3 m at a medium rate. It is not frost tender. The flowers are hermaphrodite and pollinated by insects. The plant is self-fertile. It is noted for attracting wildlife. It has a preference for light (sandy), medium (loamy), heavy (clay), and well-drained soil. It can grow in acidic, neutral, and basic (alkaline) soils, as well as in semi-shade (light woodland) or no shade. It prefers moist soil.

"Parsley" is a commonly grown culinary and medicinal herb that is often used as a domestic medicine. The fresh leaves are highly nutritious and can be considered a natural vitamin and mineral supplement in their own right. The plants prime use is as a diuretic where it is effective in ridding the body of stones and in treating jaundice, dropsy, cystitis, etc. It is also a good detoxifier, helping the body to get rid of toxins via the urine and therefore helping in the treatment of a wide range of diseases such as rheumatism. The seed is a safe herb at normal doses, but in excess it can have toxic effects. Parsley should not be used by pregnant women because it is used to stimulate menstrual flow and can therefore provoke a miscarriage. All parts of the plant can be used medicinally; the root is the part most often used though the seeds have a stronger action. Parsley is anti-dandruff, anti-spasmodic, aperient, carminative, digestive, diuretic, emmenagogue, expectorant, galactofuge, kidney, stomachic, and tonic. An infusion of the roots and seeds is taken after childbirth to promote lactation and help

Apioside: $C_{26}H_{28}O_{14}$

FIGURE 2.84 *Petroselinum crispum* (leaves and flowers) and a main component: Apioside.

contract the uterus. Parsley is also a mild laxative and is useful for treating anemia and convalescents. Caution is advised on the internal use of this herb, especially in the form of the essential oil. Excessive doses can cause liver and kidney damage, nerve inflammation, and gastro-intestinal hemorrhage. It should not be prescribed for pregnant women or people with kidney diseases. A poultice of the leaves has been applied externally to soothe bites and stings and it is also of value in treating tumors of a cancerous nature. It has been used to treat eye infections, whilst a wad of cotton soaked in the juice will relieve toothache or earache. It is also said to prevent hair loss and to make freckles disappear. If the leaves are kept close to the breasts of a nursing mother for a few days, the milk flow will cease. Parsley is said to contain the alleged "psychotroph" myristicine. Excessive contact with the plant can cause skin inflammation. Although perfectly safe to eat and nutritious in amounts that are given in recipes, parsley is toxic in excess, especially when used as an essential oil. Avoid with edema as may cause sodium and water retention. Avoid during pregnancy as parsley fruit is associated with abortions. Avoid with kidney disease. Caution with allopathic medications as associated with serotonin ($C_{10}H_{12}N_2O$) activity (Zargari, 2014; Mozaffarian, 2011; pfaf.org; Al Haadi *et al.*, 2013).

Phaseolus vulgaris (Figure 2.85) is an annual that grows up to a size of 2 m. The flowers are hermaphrodite and pollinated by bees. The plant is self-fertile. It can fix nitrogen. It has a preference for light (sandy), medium (loamy), heavy (clay), and well-drained soil. It can grow in neutral and basic (alkaline) soils. It cannot grow in the shade. It prefers moist soil.

The green pods of the "common bean" are mildly diuretic and contain a substance that reduces the blood sugar level. The dried mature pod is used, according to another report. It is used in the treatment of diabetes. The seed is diuretic, hypoglycemic, and hypotensive. Ground into a flour, it is used externally in the treatment of ulcers. The seed is also used in the treatment of cancer of the blood. When bruised and boiled with garlic the seeds have cured intractable coughs. The root is dangerously narcotic.

FIGURE 2.85   *Phaseolus vulgaris* (green pods and seeds).

A homeopathic remedy is made from the entire fresh herb. It is used in the treatment of rheumatism and arthritis, plus disorders of the urinary tract. Large quantities of the raw mature seed are poisonous. Children eating just a few seeds have shown mild forms of poisoning with nausea and diarrhea, though complete recovery took place in 12–24 hours. The toxins play a role in protecting the plant from insect predation (Zargari, 2014; Mozaffarian, 2011; pfaf.org; Yeyinou Loko *et al.*, 2018).

Pimpinella anisum (Figure 2.86) is an annual that grows to a size of 0.5 m by 0.2 m. The flowers are hermaphrodite and pollinated by insects. The plant is self-fertile. It has a preference for light (sandy), medium (loamy), and well-drained soil. It can grow in acidic, neutral, and basic (alkaline) soils. It cannot grow in the shade. It prefers dry or moist soil.

"Aniseed" has a delicious sweet licorice-like flavor and is a commonly used and very safe herbal remedy that is well suited for all age groups from children to the elderly. However, its use has declined in recent years with the advent of cheaper substitutes such as *Illicium verrum* and synthetic substances. It is a particularly useful tonic to the whole digestive system and its anti-spasmodic and expectorant effects make it of value in the treatment of various respiratory problems. The seed is the part used, generally in the form of an extracted essential oil. The essential oil comprises 70–90% anethole ($C_{10}H_{12}O$), which has an observed estrogenic effect, whilst the seed is also mildly estrogenic. This effect may substantiate the herb's use as a stimulant of sexual drive and of breast-milk production. The essential oil should not be used internally unless under professional supervision whilst the seeds are best not used

FIGURE 2.86   *Pimpinella anisum* (whole plant, seeds, and flowers).

medicinally by pregnant women, though normal culinary quantities are quite safe. The seed is antiseptic, anti-spasmodic, aromatic, carminative, digestive, expectorant, pectoral, stimulant, stomachic, and tonic. It is of great value when taken internally in the treatment of asthma, whooping cough, and pectoral affections, as well as digestive disorders such as wind, bloating, colic, nausea, and indigestion. Externally, it is used to treat infestations of lice, scabies, and as a chest rub in cases of bronchial disorders. A strong decoction of the seeds can be applied externally to swollen breasts or to stimulate the flow of milk. The German Commission E Monographs: Therapeutic guide to herbal medicines (Blumenthal *et al.*, 1998), approve *Pimpinella anisum* for coughs and bronchitis, fevers, colds, the common cold, inflammation of the mouth and pharynx, dyspepsia, and loss of appetite. Contraindicated in patients allergic to anise and anethol. Sensitization as an adverse effect observed rarely (Zargari, 2014; Mozaffarian, 2011; pfaf.org; Shojaii & Abdollahi Fard, 2012).

Pistacia vera (Figure 2.87) is a deciduous tree that grows up to a size of 10 m at a medium rate. The species is dioecious. The plant is not self-fertile. It has a preference for light (sandy), medium (loamy), and well-drained soil, and can tolerate nutritionally poor soil. It can grow in acidic, neutral, and basic (alkaline) soils, as well as very alkaline soils. It cannot grow in the shade. It prefers dry or moist soil and can tolerate drought.

The seed of the "pistachio" is rich in oil and has a pleasant mild flavor. It is very nice when eaten raw and is also widely used in confectionery, for example, ice cream, cakes, pies, etc. An edible oil is obtained from the seed but is not produced commercially due to the high price of the seed. The fruits can be made into a flavorful marmalade. Investigation into pistachio green hull has showed antioxidant, anti-microbial, and anti-mutagenic activity. A clinical trial study on young men demonstrated that a pistachio diet improved blood glucose level, endothelial function, and some indices of inflammation and oxidative status. Also, *P. vera* L. gum extract demonstrated a protective effect on oxidative damage in rat cerebral ischemia-reperfusion (Zargari, 2014; Mozaffarian, 2011; pfaf.org; Hosseinzadeh *et al.*, 2012).

Pisum sativum (Figure 2.88) is an annual that grows up to a size of 2 m. It is not frost tender. The flowers are hermaphrodite and pollinated occasionally by bees. The plant is self-fertile. It can fix nitrogen. It has a preference for light (sandy), medium (loamy), and well-drained soil. It can grow in neutral and basic (alkaline) soils. It cannot grow in the shade. It prefers moist soil.

The seed of the "pea" is contraceptive, fungistatic, and spermicidal. The dried and powdered seed has been used as a poultice on the skin where it has an appreciable affection on many types of skin complaint including acne. The oil from the seed, given once a month to women, has shown promise in preventing pregnancy by

FIGURE 2.87 *Pistacia vera* (tree, fruits, and seeds).

Legumin: $C_{21}H_{21}Cl_3Na_2O_6$

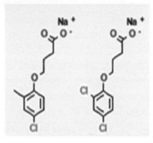

FIGURE 2.88 *Pisum sativum* (seedpods, flowers, and seeds) and its main component: Legumin.

interfering with the working of progesterone. The oil inhibits endometrial development. In trials, the oil reduced pregnancy rate in women by 60% in a two-year period
and a 50% reduction in male sperm count was achieved (Zargari, 2014; Mozaffarian,
2011; pfaf.org; Weeden *et al.*, 2018).

  Portulaca oleracea (Figure 2.89) is an annual that grows to a size of 0.3 m by 0.3 m
at a fast rate. The species is hermaphrodite and pollinated by insects. The plant is self-
fertile. It has a preference for light (sandy), medium (loamy), and well-drained soil.

Glutathione: $C_{10}H_{17}N_3O_6S$ Melatonin: $C_{13}H_{16}N_2O_2$

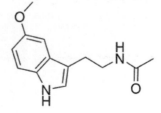

FIGURE 2.89 *Portulaca oleracea* (whole plant, flower, and leaves) and its main components:
Glutathione, melatonin, and carotene (not shown).

It can grow in acidic, neutral, and basic (alkaline) soils. It cannot grow in the shade. It prefers moist soil.

"Common purslane" is anti-bacterial, anti-scorbutic, depurative, diuretic, and febrifuge. The leaves are a rich source of omega-3 fatty acids, which is thought to be important in preventing heart attacks and strengthening the immune system. Seed sources such as walnuts, however, are much richer sources. The fresh juice is used in the treatment of strangury, coughs, sores, etc. The leaves are poulticed and applied to burns; both they and the plant juice are particularly effective in the treatment of skin diseases and insect stings. A tea made from the leaves is used in the treatment of stomach aches and headaches. The leaf juice is applied to treat earache. It is also said to alleviate caterpillar stings. The leaves can be harvested at any time before the plant flowers, and are used fresh or dried. This remedy is not given to pregnant women or to patients with digestive problems. The seeds are tonic and vermifuge. They are prescribed for dyspepsia and opacities of the cornea (Zargari, 2014; Mozaffarian, 2011; pfaf.org; Shafi & Tabassum, 2015).

Prosopis spicigera (P. cineraria) (Figure 2.90) is an evergreen tree that grows to a size of 6.5 m by 5 m at a medium rate. The flowers are pollinated by insects. It can fix nitrogen. It has a preference for light (sandy), medium (loamy), heavy (clay), and well-drained soil, and can tolerate nutritionally poor soil. It can grow in acidic, neutral, and basic (alkaline) soils. It cannot grow in the shade. It prefers dry or moist soil and can tolerate drought. The plant can tolerate strong winds but not maritime exposure.

"Jand" is reported to be astringent, demulcent, and pectoral. It is a folk remedy for various ailments. The flowers are mixed with sugar and used to prevent miscarriage. The ashes are rubbed over the skin to remove hair. The bark is considered to be anthelmintic, refrigerant, and tonic. It is used for treating asthma, bronchitis, dysentery, leukoderma, leprosy, rheumatism, muscle tremors, piles, and wandering of the mind. Smoke from the leaves is suggested for eye troubles. The pod is said to be astringent. Although recommended for scorpion stings and snake bites, the plant has not proved to be effective (Zargari, 2014; Mozaffarian, 2011; pfaf.org; Mozaffarian, 1996; Pareek *et al.*, 2015).

Piperidine: $C_5H_{11}N$

FIGURE 2.90 *Prosopis spicigera* (tree, flowers, and pods) and its main component: Piperidine.

Prunus domestica (Figure 2.91) is a deciduous tree that grows to a size of 12 m by 10 m at a medium rate. The species is hermaphrodite and pollinated by insects. The plant is self-fertile. It has a preference for light (sandy), medium (loamy), and well-drained soil, and can tolerate heavy clay soil. It can grow in acidic, neutral, and basic (alkaline) soils, as well as in semi-shade (light woodland) or no shade. It prefers moist soil.

The dried fruit of the "common plum" is a safe and effective laxative and is also stomachic. The bark is sometimes used as a febrifuge. Although no specific mention has been made of this species, all members of the genus contain amygdalin ($C_{20}H_{27}NO_{11}$) and prunasin ($C_{14}H_{17}NO_6$), substances which break down in water to form HCN. In small amounts, this exceedingly poisonous compound stimulates respiration, improves digestion, and gives a sense of well-being (Zargari, 2014; Mozaffarian, 2011; pfaf.org; Usenik *et al.*, 2013).

Prunus spinosa (Figure 2.92) is a deciduous bushy tree that grows up to a size of 3 m at a medium rate. The species is hermaphrodite and pollinated by insects. It is noted for attracting wildlife. It has a preference for light (sandy), medium (loamy), heavy (clay), and well-drained soil. It can grow in acidic, neutral, and basic (alkaline) soils, as well as in very alkaline soils. It can grow in semi-shade (light woodland) or no shade. It prefers moist soil. The plant can tolerate maritime exposure.

The flowers, bark, leaves, and fruits of "black thorn" are aperient, astringent, depurative, diaphoretic, diuretic, febrifuge, laxative, and stomachic. An infusion of the flowers is used in the treatment of diarrhea (especially for children), bladder and kidney disorders, and stomach weakness. Although no specific mention has been made of this species, all members of the genus contain amygdalin ($C_{20}H_{27}NO_{11}$) and prunasin

FIGURE 2.91   *Prunus domestica* (tree, fruits, and flowers).

FIGURE 2.92   *Prunus spinosa* (fruits, flowers, and shrub).

($C_{14}H_{17}NO_6$), substances which break down in water to form HCN. In small amounts, this exceedingly poisonous compound stimulates respiration, improves digestion, and gives a sense of well-being (Zargari, 2014; Mozaffarian, 2011; pfaf.org; Poonam *et al.*, 2011).

Punica granatum (Figure 2.93) is a deciduous tree that grows to a size of 8 m by 5 m at a medium rate. The flowers are hermaphrodite. It has a preference for light (sandy), medium (loamy), heavy (clay), and well-drained soil. It can grow in acidic, neutral, and basic (alkaline) soils. It cannot grow in the shade. It prefers dry or moist soil.

"Pomegranate" has a long history of herbal use dating back more than 3000 years. All parts of the plant contain unusual alkaloids, known as pelletierine ($C_8H_{15}NO$), which paralyze tapeworms so that they are easily expelled from the body by using a laxative. The plant is also rich in tannins, which makes it an effective astringent. It is used externally in the treatment of vaginal discharges, mouth sores, and throat infections. The whole plant, but in particular the bark, is anti-bacterial, anti-viral, and astringent. This remedy should be used with caution: Overdoses can be toxic. The flowers are used in the treatment of dysentery, stomachache, and coughs. Along with the leaves and seeds, they have been used to remove worms. The seeds are demulcent and stomachic. The fruit is a mild astringent and refrigerant in some fevers and especially in biliousness. It is also cardiac and stomachic. The dried rind of the fruit is used in the treatment of amebic dysentery, diarrhea, etc. It is a specific remedy for tapeworm infestation. The stem bark is emmenagogue. Both the stem and the root barks are used to expel tapeworms. Use this with caution, as the root bark can cause serious poisoning. The bark is harvested in the autumn and dried for later use. The dried pericarp is decocted with other herbs and used in the treatment of colic, dysentery, leucorrhoea, etc. Take recommended doses. Overdose symptoms include: Gastric irritation, vomiting, dizziness, chills, vision disorders, collapse, and death (Zargari, 2014; Mozaffarian, 2011; pfaf.org; Garachh *et al.*, 2012).

Pyrus communis (Figure 2.94) is a deciduous tree that grows up to a size of 13 m at a fast rate. The flowers are hermaphrodite and pollinated by insects. It has a preference for light (sandy), medium (loamy), and well-drained soil, and can tolerate heavy

Pelletierine: $C_8H_{15}NO$

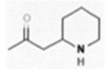

FIGURE 2.93 *Punica granatum* (tree, fruits, and flowers) and a main component: Pelletierine.

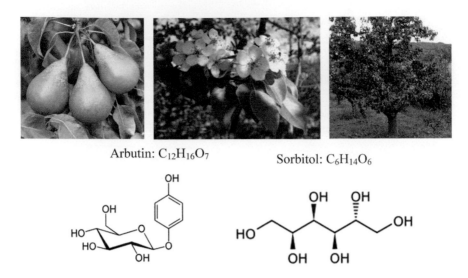

Arbutin: $C_{12}H_{16}O_7$ Sorbitol: $C_6H_{14}O_6$

FIGURE 2.94 *Pyrus communis* (fruits, flowers, and tree) and its main components: Arbutin, sorbitol.

clay soil. It can grow in acidic, neutral, and basic (alkaline) soils, as well as in semi-shade (light woodland) or no shade. It prefers moist soil and can tolerate drought. It can tolerate atmospheric pollution.

The fruit of the "European pear" is astringent, febrifuge, and sedative. The juice of the fruit is used for treating psychosis and wounds. Its leaves and bark can be used in wound healing and thus also acts as an anti-inflammatory. The leaves, buds, and bark of the tree are domestic remedies among Arabs on account of their astringent action. Pear is a rich source of vitamin C ($C_6H_8O_6$), and it is an antioxidant. Arbutin ($C_{12}H_{16}O_7$) is commonly used in urinary therapeutics and as a human skin whitening agent, as it decreases melanin in the skin. The flowers of the common pear are used in folk medicine as components of analgesic and spasmolytic drugs (Zargari, 2014; Mozaffarian, 2011; pfaf.org; Kaur & Arya, 2012).

Ranunculus sceleratus (Figure 2.95) is a perennial that grows up to a size of 0.6 m. The species is hermaphrodite and pollinated by flies. The plant is self-fertile. It has a preference for light (sandy), medium (loamy), and heavy (clay) soils. It can grow in acidic, neutral, and basic (alkaline) soils, as well as semi-shade (light woodland) or no shade. It prefers moist or wet soil and can grow in water.

"Celery-leaved buttercup" is one of the most virulent of our native plants. The whole plant is acrid, anodyne, anti-spasmodic, diaphoretic, emmenagogue, and rubefacient. When bruised and applied to the skin it raises a blister and creates a sore that is by no means easy to heal. If chewed it inflames the tongue and produces violent effects. The herb should be used fresh since it loses its effects when dried. The leaves and the root are used externally as an anti-rheumatic. The seed is tonic and is used in the treatment of colds, general debility, rheumatism, and spermatorrhea (Zargari, 2014; Mozaffarian, 2011; pfaf.org; Aslam *et al.*, 2012).

Rhaphnus sativus (Figure 2.96) is an annual that grows to a size of 0.5 m by 0.2 m at a fast rate. The flowers are hermaphrodite and pollinated by bees and flies. It has a preference for light (sandy), medium (loamy), and heavy (clay) soils. It can grow

Anemonin: $C_{10}H_8O_4$ Serotonin (5-hydroxytryptamine): $C_{10}H_{12}N_2O$

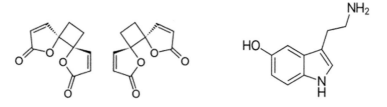

Protoanemonin (Anemonol, Ranunculol): $C_5H_4O_2$

FIGURE 2.95 *Ranunculus sceleratus* (whole plant, flowers, and leaf) and its main components: Anemonin, serotonin, protoanemonin.

FIGURE 2.96 *Raphanus sativus* (leaves, flowers, and roots).

in neutral and basic (alkaline) soils, as well as in semi-shade (light woodland) or no shade. It prefers moist soil.

The "radish" has long been grown as a food crop, but it also has various medicinal actions. The roots stimulate the appetite and digestion, having a tonic and laxative effect upon the intestines and indirectly stimulating the flow of bile. Consuming radishes generally results in improved digestion, but some people are sensitive to their acridity and robust action. The plant is used in the treatment of intestinal parasites, though the part of the plant used is not specified. The leaves, seeds, and old roots are used in the treatment of asthma and other chest complaints. The juice of the fresh leaves is diuretic and laxative. The seed is carminative, diuretic, expectorant, laxative, and stomachic; it is taken internally in the treatment of indigestion, abdominal bloating, wind, acid regurgitation, diarrhea, and bronchitis. The root is anti-scorbutic, anti-spasmodic, astringent, digestive, and diuretic. It is crushed and used as a poultice for burns, bruises, and smelly feet. Radishes are also an excellent food remedy for stone, gravel, and scorbutic conditions. The root is best harvested before the plant flowers. Its use is not recommended if the stomach or intestines are inflamed. The plant contains raphanin, which is anti-bacterial and anti-fungal. It inhibits the growth of staphylococcus aureus, E. coli, streptococci, pneumococci, etc. The plant also shows anti-tumor activity. Japanese radishes have higher concentrations of glucosinolates, a substance that acts against the thyroid gland. It is probably best to remove the skin (Zargari, 2014; Mozaffarian, 2011; pfaf.org; Kim *et al.*, 2015).

Rhus coriaria (Figure 2.97) is a deciduous bushy tree that grows up to a size of 3 m. The flowers are hermaphrodite and pollinated by bees. It has a preference for light (sandy), medium (loamy), heavy (clay), and well-drained soil. It can grow in acidic, neutral, and basic (alkaline) soils. It cannot grow in the shade. It prefers dry or moist soil and succeeds in a well-drained fertile soil in full sun.

The leaves and the seeds of the "Sicilian sumac" are astringent, diuretic, styptic, and tonic. They are used in the treatment of dysentery, hemoptysis, and conjunctivitis. The seeds are eaten before a meal in order to provoke an appetite. Some caution is advised in the use of the leaves and stems of this plant. The plant contains toxic substances which can cause severe irritation to some people. Both the sap and the fruit are poisonous (Zargari, 2014; Mozaffarian, 2011; pfaf.org; Abu-Reidah *et al.*, 2014).

Ribes rubrum (Figure 2.98) is a deciduous bushy tree that grows up to a size of 1.2 m. The species is hermaphrodite and pollinated by bees. The plant is self-fertile. It has a preference for light (sandy), medium (loamy), heavy (clay), and well-drained soil. It can grow in acidic, neutral, and basic (alkaline) soils, as well as in semi-shade (light woodland) or no shade. It prefers moist soil.

Tannic acid: $C_{76}H_{52}O_{46}$

FIGURE 2.97   *Rhus coriaria* (fruits, leaves, and shrub) and its main component: Tannic acid.

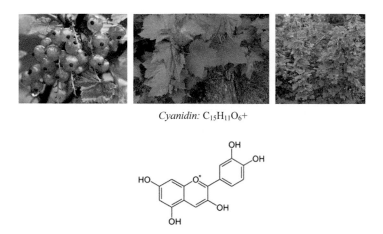

Cyanidin: $C_{15}H_{11}O_6+$

FIGURE 2.98 *Ribes rubrum* (fruits, leaves, and shrub) and its main component: Cyanidin.

The fruit of the "redcurrant" is anti-scorbutic, aperient, depurative, digestive, diuretic, laxative, refrigerant, and sialagogue. The leaves contain the toxin HCN. A concoction of the leaves is used externally to relieve rheumatic symptoms, and they are also used in poultices to relieve sprains or reduce the pain of dislocations (Zargari, 2014; Mozaffarian, 2011; pfaf.org; Kampuss & Pedersen, 2003).

Robinia pseudo-acacia (Figure 2.99) is a deciduous tree that grows to a size of 25 m by 15 m at a fast rate. The species is hermaphrodite and pollinated by bees. It can fix nitrogen. It is noted for attracting wildlife. It has a preference for light (sandy), medium (loamy), heavy (clay), and well-drained soil, and can tolerate nutritionally

Robinin: $C_{33}H_{40}O_{19}$ Benzaldehyde: C_7H_6O

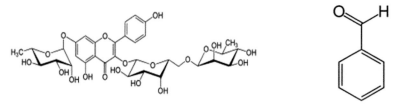

FIGURE 2.99 *Robinia pseudo-acacia* (flowers, pods, and tree) and its main components: Robinin, benzaldehyde, and kaempferol (not shown).

poor soil. It can grow in acidic, neutral, and basic (alkaline) soils. It cannot grow in the shade. It prefers dry or moist soil and can tolerate drought. It can tolerate atmospheric pollution.

The flowers of the "black locust" are anti-spasmodic, aromatic, diuretic, emollient, and laxative. They are cooked and eaten for the treatment of eye ailments. The flower is said to contain the anti-tumor compound benzaldehyde (C_7H_6O). The inner bark and the root bark are emetic, purgative, and tonic. The root bark is chewed to induce vomiting, or held in the mouth to allay toothache. The fruit is narcotic; this probably refers to the seedpod. The leaves are cholagogue and emetic and the leaf juice inhibits viruses (Zargari, 2014; Mozaffarian, 2011; pfaf.org; Patra *et al.*, 2015).

Rosa canina (Figure 2.100) is a deciduous bushy tree that grows up to a size of 3 m at a fast rate. The flowers are hermaphrodite and pollinated by bees, flies, beetles, moths, butterflies, and apomixis. The plant is self-fertile. It is noted for attracting wildlife. It has a preference for light (sandy), medium (loamy), and well-drained soil, and can tolerate heavy clay soil. It can grow in acidic, neutral, and basic (alkaline) soils, as well as in semi-shade (light woodland) or no shade. It prefers moist or wet soil. The plant can tolerate strong winds but not maritime exposure.

The petals, hips, and galls of "dog-rose" are astringent, carminative, diuretic, laxative, ophthalmic, and tonic. The hips are taken internally in the treatment of colds, influenza, minor infectious diseases, scurvy, diarrhea, and gastritis. A syrup made from the hips is used as a pleasant flavoring in medicines and is added to cough mixtures. A distilled water made from the plant is slightly astringent and is used as a lotion for delicate skin. The seeds have been used as a vermifuge. The plant is used in Bach flower remedies—the keywords for prescribing it are "resignation" and "apathy." The fruit of many members of this genus is a very rich source of vitamins and minerals, especially in vitamin C ($C_6H_8O_6$) and vitamin E ($C_{29}H_{50}O_2$), flavonoids, and other bio-active compounds. It is also a fairly good source of essential fatty acids, which is fairly unusual for a fruit. It is being investigated as a food that is capable of reducing the incidence of cancer and also as a means of halting or reversing the growth of cancers. There is a layer of hairs around the seeds just beneath the flesh of the fruit. These hairs can cause irritation to the mouth and digestive tract if ingested (Zargari, 2014; Mozaffarian, 2011; pfaf.org; Wenzig *et al.*, 2008).

Rosa damascena (Figure 2.101) is a deciduous bushy tree that grows up to a size of 1.5 m. The species is hermaphrodite and pollinated by insects. The plant is self-fertile. It has a preference for light (sandy), medium (loamy), and well-drained soil, and can tolerate heavy clay soil. It can grow in acidic, neutral, and basic (alkaline) soils, as well as in semi-shade (light woodland) or no shade. It prefers moist soil.

FIGURE 2.100 *Rosa canina* (leaves, flower, and shrub).

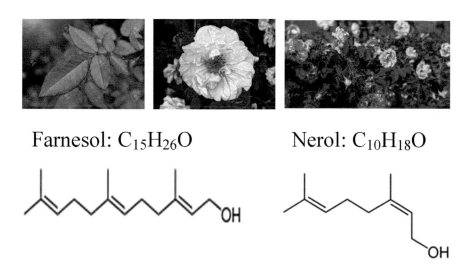

Farnesol: $C_{15}H_{26}O$ Nerol: $C_{10}H_{18}O$

FIGURE 2.101   *Rosa damascena* (leaves, flower, and shrub) and its main components: Farnesol, nerol.

The petals of the "Damask rose" are applied externally as an astringent. They are also made into a preserve and used as a tonic that helps to put on weight. The buds (the report does not say if it is leaf or flower buds) are aperient, astringent, cardiac, and tonic. They are used for removing bile and cold humors. The fruit of many members of this genus is a very rich source of vitamins and minerals, especially vitamin C ($C_6H_8O_6$) and vitamin E ($C_{29}H_{50}O_2$), flavonoids, and other bio-active compounds. It is also a fairly good source of essential fatty acids, which is fairly unusual for a fruit. It is being investigated as a food that is capable of reducing the incidence of cancer and also as a means of halting or reversing the growth of cancers (Zargari, 2014; Mozaffarian, 2011; pfaf.org; Ali *et al.*, 2016; Sarangowa *et al.*, 2014).

Rosa gallica (Figure 2.102) is a deciduous bushy tree that grows to a size of 2 m by 1 m. The flowers are hermaphrodite and pollinated by bees. It has a preference for light (sandy), medium (loamy), and well-drained soil, and can tolerate heavy clay soil. It can grow in acidic, neutral, and basic (alkaline) soils, as well as very alkaline soils.

The petals of the "French rose" are anti-bacterial, astringent, and tonic. They are taken internally in the treatment of colds, bronchial infections, gastritis, diarrhea, depression, and lethargy. Externally, they are used to treat eye infections, sore throats, minor injuries, and skin problems. The fruit of many members of this genus is a very rich source of vitamins and minerals, especially vitamin C ($C_6H_8O_6$) and vitamin E

FIGURE 2.102   *Rosa gallica* (leaves, flower, and shrub).

($C_{29}H_{50}O_2$), flavonoids, and other bio-active compounds. It is also a fairly good source of essential fatty acids, which is fairly unusual for a fruit. It is being investigated as a food that is capable of reducing the incidence of cancer and also as a means of halting or reversing the growth of cancers. The essential oil from the flowers is used in aromatherapy to counter depression, anxiety, and negative feelings. It can grow in semi-shade (light woodland) or no shade. It prefers moist soil. There is a layer of hairs around the seeds just beneath the flesh of the fruit. These hairs can cause irritation to the mouth and digestive tract if ingested (Zargari, 2014; Mozaffarian, 2011; pfaf.org; Sarangowa *et al.*, 2014).

Rubus caesius (Figure 2.103) is a deciduous bushy tree that grows to a size of 0.2 m by 1 m. The species is hermaphrodite and pollinated by bees, flies, beetles, and apomixis. The plant is self-fertile. It has a preference for light (sandy), medium (loamy), heavy (clay), and well-drained soil. It can grow in acidic, neutral, and basic (alkaline) soils, as well as in very alkaline soils. It can grow in semi-shade (light woodland) or no shade. It prefers moist soil.

The "dewberry" leaf contains ethanol (C_2H_5OH) and methanol (CH_3OH) and is used as anti-diarrheal, anti-inflammatory, anti-viral, anti-microbial, and anti-tumor, astringent, tonic, and dialysis coagulator. The dewberry fruit is low calorie and consists of vitamin C ($C_6H_8O_6$), some protein, magnesium, zinc, and copper. The fruits are used to increase the appetite. The leaves and roots can be made into tea, extracts, or an infusion to treat stomach problems such as ulcers and gastritis, and kidney stones (Zargari, 2014; Mozaffarian, 2011; pfaf.org; Schadler & Dergatschewa, 2017).

Sanguisorba minor (Figure 2.104) is an evergreen perennial that grows to a size of 0.6 m by 0.3 m. The species is hermaphrodite and pollinated by bees. The plant is self-fertile. It has a preference for light (sandy), medium (loamy), heavy (clay), and well-drained soil, and can tolerate nutritionally poor soil. It can grow in acidic, neutral, and basic (alkaline) soils, as well as very alkaline soils. It cannot grow in the shade. It prefers moist soil. The plant can tolerate maritime exposure.

Both the root and the leaves of the "salad burnet" are astringent, diaphoretic, and styptic, though the root is most active. The plant is an effective wound herb, quickly staunching any bleeding. An infusion is used in the treatment of gout and rheumatism. The leaves can be used fresh, or are harvested in July and dried (the plant should be prevented from flowering). The root is harvested in the autumn and dried. An

FIGURE 2.103   *Rubus caesius* (shrub, fruits, and flowers).

FIGURE 2.104 *Sanguisorba minor* (flower, leaves, and whole plant).

infusion of the leaves is used as a soothing treatment for sunburn or skin troubles such as eczema (Zargari, 2014; Mozaffarian, 2011; pfaf.org; Ayoub, 2003).

Sinapis alba (Figure 2.105) is an annual that grows to a size of 0.6 m by 0.3 m at a fast rate. The flowers are hermaphrodite and pollinated by bees, flies, and the wind. It has a preference for light (sandy), medium (loamy), heavy (clay), and well-drained soil. It can grow in acidic, neutral, and basic (alkaline) soils, as well as in semi-shade (light woodland) or no shade. It prefers moist soil.

The seedling plants of the "white mustard" are used in salads. The seed of the white mustard is anti-bacterial, anti-fungal, appetizer, carminative, diaphoretic, digestive, diuretic, emetic, expectorant, rubefacient, and stimulant. The seed has a cathartic action due to hydrolytic liberation of hydrogen sulfide (H_2S). In China, it is used in the treatment of coughs with profuse phlegm, tuberculosis, and pleurisy. The seed is seldom used internally as a medicine in the West. Externally, it is usually made into mustard plasters (using the ground seed), poultices, or added to bath water. It is used in the treatment of respiratory infections, arthritic joints, chilblains, and skin eruptions, etc. At a ratio of 1:3, the seed has an inhibitory action on the growth of fungus. Care should be exercised in using this remedy because the seed contains substances

Sinalbin: $C_{14}H_{19}NO_{10}S_2$

FIGURE 2.105 *Sinapis alba* (flower, leaves, and whole plant) and its main components: Sinalbin, and vitamin C (not shown).

that are extremely irritant to the skin and mucus membranes. The leaves are carminative. The seed contains substances that irritate the skin and mucous membranes. The plant is possibly poisonous once the seedpods have formed. There is the possibility of a mustard allergy especially in children and adolescents. Retention of seeds possibly in intestines if taken internally (Zargari, 2014; Mozaffarian, 2011; pfaf.org; Popova & Morra, 2014).

Sinapis arvensis (Figure 2.106) is an annual that grows to a size of 0.8 m. The species is hermaphrodite and pollinated by bees and flies. It has a preference for light (sandy), medium (loamy), and heavy (clay) soils. It can grow in acid, neutral, and basic (alkaline) soils, as well as in very alkaline soils. It cannot grow in the shade. It prefers moist soil. The plant can tolerate strong winds but not maritime exposure.

"Charlock mustard" (wild mustard, field mustard) is used in Bach flower remedies. An edible semi-drying oil is obtained from the seed. The young leaves are used as a flavoring in salads. The flowers can also be cooked as a vegetable or used as a garnish. It is good for stimulating the appetite and treatment of depression (Zargari, 2014; Mozaffarian, 2011; pfaf.org; Warwick *et al.*, 2000).

Sorbus torminalis (Figure 2.107) is a deciduous tree that grows up to a size of 20 m at a medium rate. The species is hermaphrodite and pollinated by insects. It has a preference for light (sandy), medium (loamy), and well-drained soil, and can tolerate heavy clay soil. It can grow in acidic, neutral, and basic (alkaline) soils, and can grow in semi-shade (light woodland) or no shade. It prefers moist soil. The plant can tolerate strong winds but not maritime exposure.

"Wild service tree" (chequer tree) leaves are consumed by boiling, for treatment of diabetes and stomachache. The fruits have been used for their astringent effects which are due to the high tannin content (Zargari, 2014; Mozaffarian, 2011; pfaf.org; Raudone *et al.*, 2014).

Tamarix gallica (Figure 2.108) is a deciduous bushy tree that grows up to a size of 6 m by 4 m at a medium rate. The flowers are hermaphrodite and pollinated by

gibberellic acid: $C_{19}H_{22}O_6$

FIGURE 2.106 *Sinapis arvensis* (flower, leaves, and whole plant) and its main components: Gibberellic acid, and sinalbin (not shown).

Isoquercitrin: $C_{21}H_{20}O_{12}$

FIGURE 2.107 *Sorbus torminalis* (tree, leaves, and fruits) and its main components: Isoquercitrin, and chlorogenic acid (not shown).

bees. It is noted for attracting wildlife. It has a preference for light (sandy), medium (loamy), and well-drained soil, and can tolerate heavy clay soil. It can grow in acidic, neutral, and basic (alkaline) soils, as well as very alkaline and saline soils. It cannot grow in the shade. It prefers dry or moist soil. The plant can tolerate maritime exposure.

The branchlets and the leaves of the "French tamarisk" are astringent and diuretic. An external compress is applied to wounds to stop the bleeding. The manna produced on the plant is detergent, expectorant, and laxative. Galls produced on the plant as a result of insect damage are astringent. They are used in the treatment of diarrhea and dysentery (Zargari, 2014; Mozaffarian, 2011; pfaf.org; Boulaaba *et al.*, 2015).

Tribulus terrestris (Figure 2.109) is a fast-growing annual/biennial reaching up to 0.6 m in size. It is frost tender. The flowers are hermaphrodite. It has a preference

FIGURE 2.108 *Tamarix gallica* (shrub, branchlets, and flowers).

for light (sandy), medium (loamy), heavy (clay), and well-drained soil. It can grow in acidic, neutral, and basic (alkaline) soils. It cannot grow in the shade. It prefers dry or moist soil. The plant can tolerate maritime exposure.

The seed of "Bindii" is abortifacient, alterative, anthelmintic, aphrodisiac, astringent, carminative, demulcent, diuretic, emmenagogue, galactogogue, pectoral, and tonic. It stimulates blood circulation. A decoction is used in treating impotency in males, nocturnal emissions, gonorrhea, and incontinence of urine. It has also proved effective in treating painful urination, gout, and kidney diseases. The plant has shown anti-cancer activity. The flowers are used in the treatment of leprosy. The stems are used in the treatment of scabious skin diseases and psoriasis. The dried and concocted fruits are used in the treatment of congestion, headaches, liver conditions, ophthalmia, and stomatitis (Zargari, 2014; Mozaffarian, 2011; pfaf.org; Chhatre *et al.*, 2014).

Trigonella foenom graecum (Figure 2.110) is an annual that grows to a size of 0.6 m by 0.4 m at a fast rate. The flowers are hermaphrodite and pollinated by insects. It can fix nitrogen. It has a preference for light (sandy), medium (loamy), and heavy (clay) soils. It can grow in acidic, neutral, and basic (alkaline) soils. It cannot grow in the shade. It prefers dry or moist soil.

"Fenugreek" is much used in herbal medicine, especially in North Africa, the Middle East, and India. It has a wide range of medicinal applications. The seeds are very nourishing and are given to convalescents and to encourage weight gain, especially in anorexia nervosa. The seeds should not be prescribed medicinally for pregnant women as they can induce uterine contractions. Research has shown that the seeds can inhibit cancer of the liver, lower blood cholesterol levels, and also have an anti-diabetic effect. The seeds and leaves are anti-cholesterolemic, anti-inflammatory, anti-tumor, carminative, demulcent, deobstruent, emollient, expectorant, febrifuge, galactogogue, hypoglycemic, laxative, parasiticide, restorative, and uterine tonic. The seed yields a strong mucilage and is therefore useful in the treatment of inflammation and ulcers of the stomach and intestines. Taken internally, a decoction of the ground

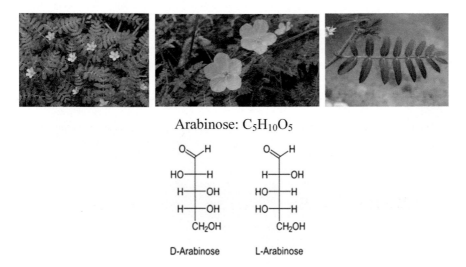

Arabinose: $C_5H_{10}O_5$

D-Arabinose   L-Arabinose

FIGURE 2.109 *Tribulus terrestris* (whole plant, flowers, and leaf) and its main component: Arabinose.

Trigonelline: $C_7H_7NO_2$

FIGURE 2.110 *Trigonella foenum graecum* (leaves, seeds, and flower) and its main components: Trigonelline, and vitamin B_3 (not shown).

seeds serves to drain off the sweat ducts. The seed is very nourishing and body build-ing and is one of the most efficacious tonics in cases of physical debility caused by anemia or by infectious diseases, especially where a nervous factor is involved. It is also used in the treatment of late-onset diabetes, poor digestion (especially in con-valescence), insufficient lactation, painful menstruation, labor pains, etc. The seeds freshen bad breath and restore a dulled sense of taste. Externally, the seeds can be ground into a powder and used as a poultice for abscesses, boils, ulcers, burns, etc., or they can be used as a douche for excessive vaginal discharge. The leaves are harvested in the growing season and can be used fresh or dried. The seeds are harvested when fully ripe and dried for later use. Compounds extracted from the plant have shown cardiotonic, hypoglycemic, diuretic, anti-phlogistic, and hypotensive activity. One of its constituent alkaloids, called "trigonelline," has shown potential for use in cancer therapy. The seed contains the saponin diosgenin, an important substance in the syn-thesis of oral contraceptives and sex hormones, whilst saponins in the plant have been extracted for use in various other pharmaceutical products. The German Commission E Monographs: Therapeutic guide to herbal medicines (Blumenthal *et al.*, 1998), approve *Trigonella foenum graecum* for loss of appetite, inflammation of the skin. The seed contains 1% saponins. Although poisonous, saponins are poorly absorbed by the human body and so most pass through without harm. Saponins are quite bitter and can be found in many common foods such as some beans. They can be removed by carefully leaching the seed or flour in running water. Thorough cooking, and per-haps changing the cooking water once, will also remove most of them. However, it is not advisable to eat large quantities of food that contain saponins. Saponins are much more toxic to some creatures, such as fish, and hunting tribes have traditionally put large quantities of them in streams, lakes, etc., in order to stupefy or kill the fish.

Care for diabetics on anti-diabetic allopathic as may lower blood sugar. Can affect drug absorption as high fiber content. Constituents can alter the effects of monoamine oxide inhibitors (Zargari, 2014; Mozaffarian, 2011; pfaf.org; Moradi Kor *et al.*, 2013).

Tropaeolum majus (Figure 2.111) is a perennial climber that grows to a size of 3.5 m by 1.5 m at a fast rate. The species is hermaphrodite. It has a preference for light (sandy), medium (loamy), and well-drained soil, and can tolerate nutritionally poor soil. It can grow in acidic, neutral, and basic (alkaline) soils. It cannot grow in the shade. It prefers moist soil.

"Nasturtium" has long been used in Andean herbal medicine as a disinfectant and wound-healing herb, and as an expectorant to relieve chest conditions. All parts of the plant appear to be antibiotic and an infusion of the leaves can be used to increase resistance to bacterial infections and to clear nasal and bronchial catarrh. The remedy seems to both reduce catarrh formation and stimulate the clearing and coughing up of phlegm. The leaves are anti-bacterial, anti-fungal, antiseptic, aperient, depurative, diuretic, emmenagogue, expectorant, laxative, and stimulant. A glycoside found in the plant reacts with water to produce an antibiotic. The plant has antibiotic properties towards aerobic spore forming bacteria. Extracts from the plant have anti-cancer activity. The plant is taken internally in the treatment of genito-urinary diseases, respiratory infections, scurvy, and poor skin and hair conditions. Externally, it makes an effective antiseptic wash and is used in the treatment of baldness, minor injuries, and skin eruptions. Any part of the plant can be used; it is harvested during the growing season and used fresh (Zargari, 2014; Mozaffarian, 2011; pfaf.org; Carvalho *et al.*, 2015).

Vicia sativa (Figure 2.112) is an annual climber that grows to a size of 1.2 m at a fast rate. It is not frost tender. The flowers are hermaphrodite and pollinated by

Glucotropaeolin: $C_{14}H_{19}NO9S_2$

FIGURE 2.111 *Tropaeolum majus* (whole plant, flower, and leaves) and its main component: Glucotropaeolin.

bees. The plant is self-fertile. It can fix nitrogen. It has a preference for light (sandy), medium (loamy), heavy (clay), and well-drained soil. It can grow in acidic, neutral, and basic (alkaline) soils, as well as in semi-shade (light woodland) or no shade. It prefers moist soil.

"Common vetch" is useful for treating measles, smallpox, and inflammation. There is some evidence that the seed may be toxic, but this has only been shown under laboratory conditions. There are no recorded cases of poisoning by this plant (Zargari, 2014; Mozaffarian, 2011; pfaf.org; Schlereth *et al.*, 2000).

Viola odorata (Figure 2.113) is an evergreen perennial that grows to a size of 0.1 m by 0.5 m at a fast rate. It is not frost tender. The flowers are hermaphrodite and pollinated by bees and cleistogamous. The plant is self-fertile. It has a preference for light (sandy), medium (loamy), heavy (clay), and well-drained soil. It can grow in acidic, neutral, and basic (alkaline) soils, as well as in semi-shade (light woodland) or no shade. It prefers moist soil.

"Sweet violet" has a long and proven history of folk use, especially in the treatment of cancer and whooping cough. It also contains salicylic acid ($C_7H_6O_3$), which is

Vicine: $C_{10}H_{16}N_4O_7$

FIGURE 2.112 *Vicia sativa* (whole plant, flower, and leaves) and its main components: Vicine, and legumin (not shown).

FIGURE 2.113 *Viola odorata* (whole plant, flower, and leaf).

used to make aspirin. It is therefore effective in the treatment of headaches, migraine, and insomnia. The whole plant is anti-inflammatory, diaphoretic, diuretic, emollient, expectorant, and laxative. It is taken internally in the treatment of bronchitis, respiratory catarrh, coughs, asthma, and cancer of the breast, lungs, or digestive tract. Externally, it is used to treat mouth and throat infections. The plant can either be used fresh, or harvested when it comes into flower and then be dried for later use. The flowers are demulcent and emollient, and are used in the treatment of biliousness and lung troubles. The petals are made into a syrup and used in the treatment of infantile disorders. The roots are a much stronger expectorant than other parts of the plant, but they also contain the alkaloids which at higher doses are strongly emetic and purgative. They are gathered in the autumn and dried for later use. The seeds are diuretic and purgative. They have been used in the treatment of urinary complaints and are considered a good remedy for gravel. A homeopathic remedy is made from the whole fresh plant. It is considered useful in the treatment of spasmodic coughs and rheumatism of the wrist. An essential oil from the flowers is used in aromatherapy in the treatment of bronchial complaints, exhaustion, and skin complaints. It may cause vomiting. Possible additive effect with laxatives (Zargari, 2014; Mozaffarian, 2011; pfaf.org; Mittal *et al.*, 2015).

Vitis vinifera (Figure 2.114) is a fast-growing deciduous climber reaching up to 15 m in size at a fast rate. It is not frost tender. The flowers are hermaphrodite and pollinated by insects. It has a preference for light (sandy), medium (loamy), heavy (clay), and well-drained soil. It can grow in acidic, neutral, and basic (alkaline) soils, as well as in semi-shade (light woodland) or no shade. It prefers dry or moist soil.

The "grape" is a nourishing and slightly laxative fruit that can support the body through illness, especially of the gastro-intestinal tract and liver. The fresh fruit is anti-lithic, constructive, cooling, diuretic, and strengthening. A period of time on a diet based entirely on the fruit is especially recommended in the treatment of torpid liver or sluggish biliary function. The fruit is also helpful in the treatment of varicose veins, hemorrhoids, and capillary fragility. The dried fruit is demulcent, cooling, mildly expectorant, laxative, and stomachic. It has a slight effect in easing coughs. The leaves, especially red leaves, are anti-inflammatory and astringent. A decoction

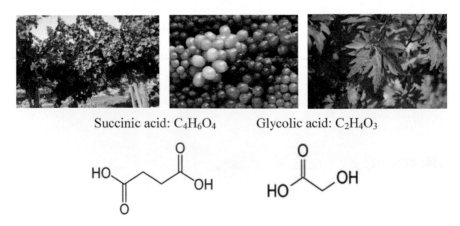

Succinic acid: $C_4H_6O_4$ Glycolic acid: $C_2H_4O_3$

FIGURE 2.114   *Vitis vinifera* (shrub, fruits, and leaves) and its main components: Succinic acid, glycolic acid.

is used in the treatment of threatened abortion, internal and external bleeding, cholera, dropsy, diarrhea, and nausea. It is also used as a wash for mouth ulcers and as douche for treating vaginal discharge. Red grape leaves are also helpful in the treatment of varicose veins, hemorrhoids, and capillary fragility. The leaves are harvested in early summer and used fresh or dried. The seed is anti-inflammatory and astringent. The sap of young branches is diuretic. It is used as a remedy for skin diseases and is also an excellent lotion for the eyes. The tendrils are astringent, and a decoction is used in the treatment of diarrhea. The plant is used in Bach flower remedies—the keywords for prescribing it are "dominating," "inflexible," and "ambitious" (Zargari, 2014; Mozaffarian, 2011; pfaf.org; Zheng Feei *et al.*, 2017).

Ziziphus jujuba (Figure 2.115) is a deciduous tree that grows to a size of 10 m by 7 m at a fast rate. The flowers are hermaphrodite and pollinated by insects. The plant is self-fertile. It has a preference for light (sandy), medium (loamy), heavy (clay), and well-drained soil, and can tolerate nutritionally poor soil. It can grow in acidic, neutral, and basic (alkaline) soils, as well as very alkaline soils. It cannot grow in the shade. It prefers dry or moist soil and can tolerate drought.

"Jujuba" is both a delicious fruit and an effective herbal remedy. It aids weight gain, improves muscular strength, and increases stamina. In Chinese medicine, it is prescribed as a tonic to strengthen liver function. Japanese research has shown that jujuba increases immune-system resistance. In one clinical trial in China, 12 patients with liver complaints were given jujuba, peanuts, and brown sugar nightly. In four weeks, their liver function had improved. It is antidote, diuretic, emollient, and expectorant. The dried fruits contain saponins, triterpenoids, and alkaloids. They are anodyne, anti-cancer, pectoral, refrigerant, sedative, stomachic, styptic, and tonic. They are considered to purify the blood and aid digestion. They are used internally in the treatment of a range of conditions including chronic fatigue, loss of appetite, diarrhea, pharyngitis, bronchitis, anemia, irritability, and hysteria. The seed contains a number of medically active compounds including saponins, triterpenes, flavonoids,

Vitamin B_1 (Thiamine, Thiamin): $C_{12}H_{17}N_4OS+$

FIGURE 2.115 *Ziziphus jujuba* (leaves, fruits, and tree) and a main component: Vitamins B_1.

and alkaloids. It is hypnotic, narcotic, sedative, stomachic, and tonic. It is used internally in the treatment of palpitations, insomnia, nervous exhaustion, night sweats, and excessive perspiration. The root is used in the treatment of dyspepsia. A decoction of the root has been used in the treatment of fevers. The root is made into a powder and applied to old wounds and ulcers. The leaves are astringent and febrifuge. They are said to promote the growth of hair. They are used to form a plaster in the treatment of strangury. The plant is a folk remedy for anemia, hypertonia, nephritis, and nervous diseases. The plant is widely used in China as a treatment for burns. Caution for diabetics on allopathic medication (Zargari, 2014; Mozaffarian, 2011; pfaf.org; Cui *et al.*, 2017).

3

Spermatophytes: Angiosperms, Dicotyledons, Sympetalaes

Spermatophytes are split into several sections. This chapter reviews angiosperms, dicotyledons, and sympetalaes which are listed in Table 3.1. A total of 21 species is provided and described in detail below. As each species is presented, information on the taxonomy, plant use, and specifically the key natural products isolated from each plant is provided.

In this group, the flowers are usually bisexual and contain sepals and petals. The petals are continuous. Another name for this group is gamopetales (Azadbakht, 2000).

Diversity of this group in Iran is low, but it consists of two important families; caprifoliaceae and labiateae, which have important uses.

Galium verum (Figure 3.1) is a perennial that grows to a size of 0.6 m by 1 m. The species is hermaphrodite and is pollinated by flies and beetles. The plant is self-fertile. It is noted for attracting wildlife. Preference for light (sandy), medium (loamy), and heavy (clay) soils, and well-drained soil. It can grow in acidic, neutral, and basic (alkaline) soils, also semi-shade (light woodland) or no shade. It prefers dry or moist soil. The plant can tolerate maritime exposure.

"Lady's bedstraw" has a long history of use as a herbal medicine, though it is little used in modern medicine. Its main application is as a diuretic and treatment for skin complaints. The leaves, stems, and flowering shoots are anti-spasmodic, astringent, diuretic, used in foot care, lithontripic, and vulnerary. The plant is used as a remedy in gravel, stone, or urinary disorders and is believed to be a remedy for epilepsy. A powder made from the fresh plant is used to soothe reddened skin and reduce inflammation whilst the plant is also used as a poultice on cuts, skin infections, slow-healing wounds, etc. The plant is harvested as it comes into flower and is dried for later use. Both asperuloside ($C_{18}H_{22}O_{11}$) and coumarin ($C_9H_6O_2$) occur in some species of galium. Asperuloside ($C_{18}H_{22}O_{11}$) can be converted into prostaglandins (hormone-like compounds that stimulate the uterus and affect blood vessels), making the genus of great interest to the pharmaceutical industry (Zargari, 2014; Mozaffarian, 2011; pfaf.org; Bradic et al., 2017).

Gardenia jasminoides (Figure 3.2) is a bushy tree with greyish bark and dark green, shiny, evergreen leaves with prominent veins. The white flowers have a matte texture, in contrast to the glossy leaves. They can be quite large, up to 10 cm in diameter, loosely funnel-shaped, and double in form. Blooming in summer and autumn, they are among the most strongly fragrant of all flowers. They are followed by small, oval fruit.

The iridoids, genipin ($C_{11}H_{14}O_5$), and geniposidic acid ($C_{16}H_{22}O_{10}$) can be found in "Cape jasmine." Crocetin ($C_{20}H_{24}O_4$) (a chemical compound usually obtained from *Crocus sativus*) can also be obtained from the fruit of *G. jasminoides*. Industrial use

TABLE 3.1

Spermatophytes: Angiosperms, Dicotyledons, Sympetalaes

Family Name	Species Name	Major Features of Botany	Usable Parts	Some Important Substances	Edible, Some Industrial and Medicinal Benefits
Caprifoliaceae	*Lonicera caprifolium*	Climber—White to Yellow Flower—Dark Green Leaf	Leaf—Flower—Seed—Fruit	Melitane—Cholorgenic Acid	Heart Tonic—Astringent—Emetic—Cathartic—Emollient—Expectorant—Diuretic—Anti-Cough and Spasmodic
Caprifoliaceae	*Sambucus ebulus*	Herbal—White Flower	Root—Leaf—Fruit	Flavonoids—Saponins—Malic Acid—Pectin	Diuretic—Sweaty—Aperient—Cholagogue—Diaphoretic—Expectorant—Laxative—Treating Swellings and Contusions—Anti-Phlogistic
Caprifoliaceae	*Sambucus nigra*	Bushy Tree—Dark Green Leaf	Inner Bark—Flower—Fruit—Leaf	Sambucine—Sambunigrin	Sweaty—Laxative—Diuretic—Expectorant Hemostatic—Emetic—Depurative—Skin Hygiene—Purgative—Treating Rheumatism, Constipation, and Arthritic Conditions
Caprifoliaceae	*Sambucus racemosa*	Bushy Tree—White Flower	Leaf—Blossom—Bark of Stem	Sucrose—Glucoside	Treating Measles—Sweaty—Diuretic—Purgative
Caprifoliaceae	*Viburnum lantana*	Bushy Tree—White Flower—Thick Leaf with Trichome	Leaf—Fruit—Bark of Stem	Triterpenoids—Diterpenoids—Polyphenols	Treating Diarrhea—Sore Throat—Gum Inflammation
Caprifoliaceae	*Viburnum opulus*	Bushy Tree—White Flower—Red Fruit	Bark of Stem—Leaf—Fruit	Viburnine—Scopoletin—Vitamins C and K	Treating Paroxysm—Anti-Spasmodic and Scorbutic—Astringent—Diuretic—Sedative—Emetic—Laxative
Labiateae	*Lamium album*	Herbal—White Flower—Heart-Like Leaf	Leaf—Flower	Tannins—Mucilage—Gallic Acid	Treating Anemia and Respiratory Diseases—Astringent—Demulcent Cholagogue—Depurative—Diuretic—Expectorant—Hemostatic—Hypnotic—Pectoral—Resolvent—Sedative—Styptic—Tonic—Anti-Spasmodic

(Continued)

TABLE 3.1 (CONTINUED)

Spermatophytes: Angiosperms, Dicotyledons, Sympetalaes

Family Name	Species Name	Major Features of Botany	Usable Parts	Some Important Substances	Edible, Some Industrial and Medicinal Benefits
Labiateae	*Melissa officinalis*	Herbal—Sharp-Tip Leaf—White Flower	Young Shoot—Leaf	Citral	Treating Paroxysm and Dizziness —Carminative—Diaphoretic— Digestive—Emmenagogue—Febrifuge— Sedative—Tonic—Anti-Bacterial, Anti-Spasmodic, and Anti-Viral
Labiateae	*Mentha piperita*	Herbal—With Rhizome and Stolon	Herb—Leaf	Tannins—Menthol— Isomenthol	Edible—Treating Bloating—Abortifacient— Anodyne—Carminative—Cholagogue— Diaphoretic—Refrigerant—Stomachic— Tonic—Vasodilator—Antiseptic, Anti-Bacterial, and Anti-Spasmodic
Labiateae	*Mentha pulegium*	Herbal—Cylindrical Appearance	Herb—Leaf	Pectin—Resin— Menthol—Pulegone— Vitamins A and E	Edible—Treating Bloating, Itchiness, and Formication Disinfectant—Carminative—Diaphoretic— Emmenagogue—Sedative—Stimulant— Antiseptic and Anti-Spasmodic
Labiateae	*Nepeta cataria*	Herbal—With Trichome—White Flower—With Good Smell	Herb—Leaf—Flower— Stem	Nepetalactone	Treating Paroxysm, Fevers, and Colds—Anti- Spasmodic and Anti-Tussive—Astringent—Carminative— Diaphoretic—Emmenagogue— Refrigerant—Sedative—Stimulant— Stomachic—Tonic

(Continued)

Natural Products and Botanical Medicines of Iran

TABLE 3.1 (CONTINUED)

Spermatophytes: Angiosperms, Dicotyledons, Sympetalaes

Family Name	Species Name	Major Features of Botany	Usable Parts	Some Important Substances	Edible, Some Industrial and Medicinal Benefits
Labiateae	*Ocimum basilicum*	Herbal—Sharp-Tip Leaf—White Flower	Leaf—Flower—Seed	Estragole—O-cymene	Edible—Treating Dizziness, Nausea, and Coughs
Labiateae	*Origanum majorana*	Herbal—Small White Flower	Flowered Shoot—Leaf—Herb	Terpineol—Terpinene—Sabinene	Tonic—Diuretic—Sedative—Carminative—Cholagogue—Diaphoretic—Emmenagogue—Expectorant—Stimulant—Stomachic—Antiseptic and Anti-Spasmodic
Labiateae	*Satureja hortensis*	Herbal—Stem Color Darker Than Leaves	Herb	Ursolic Acid	Edible—Antiseptic—Aromatic—Carminative—Digestive—Expectorant—Stomachic Diuretic—Treating Bloating, Nausea, Diarrhea, Bronchial Congestion, Sore Throat, and Menstrual Disorders
Plantagiaceae	*Plantago major*	Herbal—Nervure Throughout the Leaf	Leaf—Seed	Mucilage—Aucubin	Edible—Treating Diarrhea, Asthma, Gastritis, Peptic Ulcers, Irritable Bowel Syndrome, Hemorrhage, Hemorrhoids, Cystitis, Bronchitis, Catarrh, Sinusitis, and Hayfever—Dialysis—Astringent—Demulcent—Deobstruent—Depurative—Diuretic—Expectorant—Hemostatic—Refrigerant
Plantaginaceae	*Plantago psyllium*	Herbal—Narrow Leaf—Small Flower	Herb—Seed	Aucubin—Xylene—Mucilage—Arabinose	Edible—Laxative—Wound Healing—Demulcent—Emollient

(Continued)

TABLE 3.1 (CONTINUED)

Spermatophytes: Angiosperms, Dicotyledons, Sympetalaes

Family Name	Species Name	Major Features of Botany	Usable Parts	Some Important Substances	Edible, Some Industrial and Medicinal Benefits
Plumbaginaceae	*Plumbago europaea*	Herbal—Leaves Aspheric at Base, Other Leaves Sharp Tip and Narrow—Light Blue Flower	Herb—Root	Tannins—Starch—Plumbagin—Gallic Acid	Aperient—Emetic—Irritating—Acrid, Sialagogue—Vesicant— Treating Obesity and Toothache
Rubiaceae	*Galium verum*	Herbal—Narrow Leaf—Small and Yellow Leaf	Flowered Shoot—Leaf—Stem	Gallotannic Acid—Citric Acid—Coumarin—Asperuloside	Industrial (Cheese Making and Wool Coloring)—Astringent—Diuretic—Sedative—Lithontripic—Vulnerary—Anti-Spasmodic and Anti-Inflammatory
Rubiaceae	*Gardenia jasminoides*	Bushy Tree—White and Aromatic Flower—Evergreen	Flower—Fruit	Geniposidic Acid—Crocetin—Genipin	Industrial (Tea Fragrant, Dye Making)—Treating Arthritis and Inflammation
Rubiaceae	*Rubia tinctorum*	Herbal—Spiny Stem—Small Yellow Flower	Root—Rhizome	Alizarin—Rubiadin—Purpurin—Ruberythric Acid	Industrial (Fabric Coloring)—Diuretic—Tonic—Laxative—Aperient—Astringent—Cholagogue—Emmenagogue— Treating Amenorrhea, Dropsy, and Jaundice
Valerianaceae	*Valeriana officinalis*	Herbal—Aromatic White or Pink Flower—With Rhizome—Plenty of Roots	Root—Rhizome	Tannins—Valerenic Acid	Treating Bloating, Epilepsy, Cramps, Hypertension, and Paroxysm—Anti-Spasmodic—Carminative—Diuretic—Hypnotic—Sedative—Stimulant

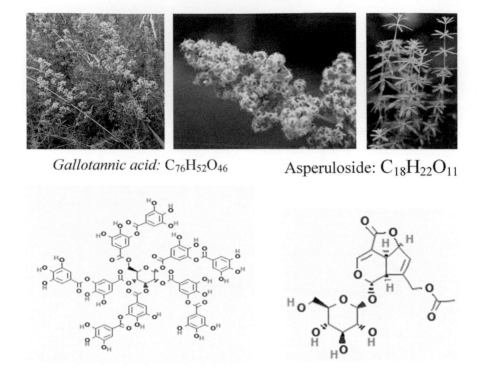

Gallotannic acid: C₇₆H₅₂O₄₆   Asperuloside: C₁₈H₂₂O₁₁

FIGURE 3.1   *Galium verum* (whole plant, flowers, and leaves) and its main components: Gallotannic acid, asperuloside.

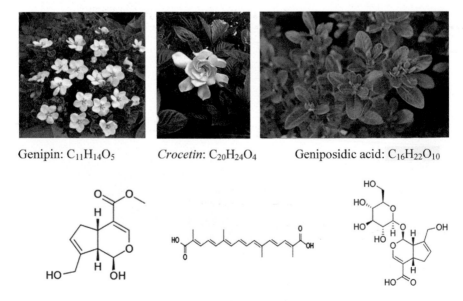

Genipin: C₁₁H₁₄O₅   *Crocetin:* C₂₀H₂₄O₄   Geniposidic acid: C₁₆H₂₂O₁₀

FIGURE 3.2   *Gardenia jasminoides* (shrub, flower, and leaves) and its main components: Genipin, crocetin, geniposidic acid.

of *G. jasminoides* is for fragrant tea. It is useful for treating inflammation and arthritis (Zargari, 2014; Mozaffarian, 2011; pfaf.org; Thantsin, 2011; Xiao *et al.*, 2016).

Lamium album (Figure 3.3) is a perennial that grows to a size of 0.6 m by 1 m at a medium rate. The species is hermaphrodite and is pollinated by bees. It is noted for attracting wildlife. Preference for light (sandy), medium (loamy), and heavy (clay) soils, also well-drained soil and can tolerate heavy clay soil. It can grow in acids, neutral, and basic (alkaline) soils, also semi-shade (light woodland) or no shade. It prefers moist soil.

"White dead nettle" is an astringent and demulcent herb that is chiefly used as a uterine tonic, to arrest inter-menstrual bleeding and to reduce excessive menstrual flow. It is a traditional treatment for abnormal vaginal discharge and is sometimes taken to relieve painful periods. The flowering tops are anti-spasmodic, astringent, cholagogue, depurative, diuretic, expectorant, hemostatic, hypnotic, pectoral, resolvent, sedative, styptic, tonic, vasoconstrictor, and vulnerary. An infusion is used in the treatment of kidney and bladder complaints, diarrhea, menstrual problems, bleeding after childbirth, vaginal discharges, and prostatitis. Externally, the plant is made into compresses and applied to piles, varicose veins, and vaginal discharges. A distilled water from the flowers and leaves makes an excellent and effective eye lotion to relieve ophthalmic conditions. The plant is harvested in the summer and can be dried for later use. A homeopathic remedy is made from the plant. It is used in the treatment of bladder and kidney disorders and amenorrhea (Zargari, 2014; Mozaffarian, 2011; pfaf.org; Czerwinska *et al.*, 2018).

Lonicera caprifolium (Figure 3.4) is a deciduous climber that grows to a size of 6 m by 6 m at a medium rate. The species is hermaphrodite and is pollinated by moths and butterflies. Preference for light (sandy), medium (loamy), and heavy (clay) soils.

FIGURE 3.3 *Lamium album* (whole plant, flowers, and leaves).

FIGURE 3.4 *Lonicera caprifolium* (whole plant, flowers, and leaves).

It can grow in acidic, neutral, and basic (alkaline) soils, also full shade (deep wood-land), semi-shade (light woodland), or no shade. It prefers moist soil.

The fruit of "Italian woodbine" is emetic and cathartic. The pressed juice makes a mild purgative. The leaves and flowers are anti-spasmodic, emollient, and expectorant. They are used as a cutaneous and mucous tonic and as a vulnerary. Recent research has shown that the plant has an outstanding curative action in cases of colitis. The seed is diuretic (Zargari, 2014; Mozaffarian, 2011; pfaf.org; Bergqvist & Olsson, 1992).

Melissa officinalis (Figure 3.5) is a perennial that grows to a size of 0.7 m by 0.4 m at a fast rate. It is not frost tender. The flowers are hermaphrodite and are pollinated by bees. It is noted for attracting wildlife. Preference for light (sandy) and medium (loamy) soils, and well-drained soil. It can grow in acid, neutral, and basic (alkaline) soils, also semi-shade (light woodland) or no shade. It prefers dry or moist soil and can tolerate drought.

"Lemon balm" is a commonly grown household remedy with a long tradition as a tonic remedy that raises the spirits and lifts the heart. Modern research has shown that it can help significantly in the treatment of cold sores. The leaves and young flow-ering shoots are anti-bacterial, anti-spasmodic, anti-viral, carminative, diaphoretic, digestive, emmenagogue, febrifuge, sedative, and tonic. It also acts to inhibit thyroid activity. An infusion of the leaves is used in the treatment of fevers and colds, indi-gestion associated with nervous tension, excitability, and digestive upsets in children, hyperthyroidism, depression, mild insomnia, headaches, etc. Externally, it is used to treat herpes, sores, gout, insect bites, and as an insect repellent. The plant can be used fresh or dried; for drying it is harvested just before or just after flowering. The essential oil contains citral ($C_{10}H_{16}O$), which acts to calm the central nervous sys-tem and is strongly anti-spasmodic. The plant also contains polyphenols, believed to combat the herpes simplex virus which produces cold sores. The essential oil is used in aromatherapy. Its keyword is "female aspects." It is used to relax and rejuvenate, especially in cases of depression and nervous tension. The German Commission E Monographs state: Therapeutic guide to herbal medicines (Blumenthal *et al.*, 1998), approve *Melissa officinalis* for nervousness and insomnia. It can cause irritation in high concentrates, avoid during pregnancy and care with sensitive skin (Zargari, 2014; Mozaffarian, 2011; pfaf.org; Moradpour *et al.*, 2017).

Mentha piperita (Figure 3.6) is a perennial that grows to a size of 1 m by 0.5 m. It is not frost tender. The flowers are hermaphrodite and are pollinated by insects. It is noted for attracting wildlife. Preference for light (sandy), medium (loamy), and heavy (clay) soils, and can tolerate heavy clay soil. It can grow in acidic, neutral, and basic (alkaline) soils and semi-shade (light woodland) or no shade. It prefers moist soil.

FIGURE 3.5 *Melissa officinalis* (whole plant, flowers, and leaves).

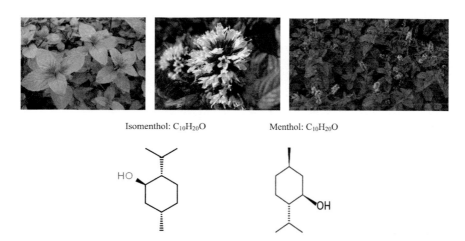

Isomenthol: C$_{10}$H$_{20}$O   Menthol: C$_{10}$H$_{20}$O

FIGURE 3.6 *Mentha piperita* (leaves, flowers, and whole plant) and its main components: Isomenthol, menthol.

"White peppermint" is a very important and commonly used remedy, being employed by allopathic doctors as well as herbalists. It is also widely used as a domestic remedy. This cultivar is milder acting than black peppermint (*Mentha x piperita vulgaris*). A tea made from the leaves has traditionally been used in the treatment of fevers, headaches, digestive disorders (especially flatulence), and various minor ailments. The herb is abortifacient, anodyne, antiseptic, anti-spasmodic, carminative, cholagogue, diaphoretic, refrigerant, stomachic, tonic, and vasodilator. An infusion is used in the treatment of irritable bowel syndrome, digestive problems, spastic colon, etc. Externally a lotion is applied to the skin to relieve pain and reduce sensitivity. The leaves and stems can be used fresh or dried; they are harvested for drying in August as the flowers start to open. The essential oil in the leaves is antiseptic and strongly antibacterial, though it is toxic in large doses. When diluted it can be used as an inhalant and chest rub for respiratory infections. The essential oil is used in aromatherapy. Its keyword is "cooling." In large quantities, this plant, especially in the form of the extracted essential oil, can cause abortions so should not be used by pregnant women (Zargari, 2014; Mozaffarian, 2011; pfaf.org; Patil *et al.*, 2016).

Mentha pulegium (Figure 3.7) is a perennial that grows to a size of 0.6 m by 0.4 m. It is not frost tender. The flowers are hermaphrodite and are pollinated by bees. It is noted for attracting wildlife. Preference for light (sandy), medium (loamy), and heavy (clay) soils and can tolerate heavy clay soil. It can grow in acidic, neutral, and basic (alkaline) soils, also semi-shade (light woodland) or no shade. It prefers moist soil.

"Pennyroyal" has been used for centuries in herbal medicine. Its main value is as a digestive tonic where it increases the secretion of digestive juices and relieves flatulence and colic. Pennyroyal also powerfully stimulates the uterine muscles and encourages menstruation; thus, it should not be prescribed for pregnant women as it can procure abortions; this is especially the case if the essential oil is used. The herb is antiseptic, anti-spasmodic, carminative, diaphoretic, emmenagogue, sedative, and stimulant. A tea made from the leaves has traditionally been in the treatment of fevers, headaches, minor respiratory infections, digestive disorders, menstrual complaints, and various minor ailments. It is occasionally used as a treatment for

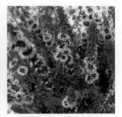

Pulegone: C₁₀H₁₆O

Pulegone: $C_{10}H_{16}O$

FIGURE 3.7   *Mentha pulegium* (leaves, flower, and whole plant) and a main components: Pulegone.

intestinal worms. Externally, an infusion is used to treat inflamed skin disorders such as eczema and rheumatic conditions such as gout. The leaves are harvested in the summer as the plant comes into flower and are dried for later use. The essential oil in the leaves is antiseptic, though it is toxic in large doses. In large quantities, this plant, especially in the form of the extracted essential oil, can cause abortions so it should not be used by pregnant women. Avoid if the patient has fits or seizures and those with liver or kidney disease. Oral intake may cause abdominal cramps, fever, nausea, vomiting, confusion, delirium, auditory and visual hallucinations (Zargari, 2014; Mozaffarian, 2011; pfaf.org; Politeo *et al.*, 2018).

Nepeta cataria (Figure 3.8) is a perennial that grows to a size of 1 m by 0.6 m. The species is hermaphrodite and is pollinated by bees. It is noted for attracting wildlife. Preference for light (sandy) and medium (loamy) soils, and well-drained soil. It can grow in acidic, neutral, and basic (alkaline) soils, and very alkaline soils. It cannot grow in the shade. It prefers dry or moist soil.

"Catmint" (catnip), has a long history of use as a household herbal remedy, being employed especially in treating disorders of the digestive system and, as it stimulates sweating, it is useful in reducing fevers. The herb's pleasant taste and gentle action make it suitable for treating colds, flu, and fevers in children. It is more effective when used in conjunction with elder flower (*Sambucus nigra*). The leaves and flowering tops are strongly anti-spasmodic, anti-tussive, astringent, carminative, diaphoretic, slightly emmenagogue, refrigerant, sedative, slightly stimulant, stomachic, and tonic. The flowering stems are harvested in August when the plant is in full flower; they are dried and stored for use as required. An infusion produces free perspiration, which is beneficial in the treatment of fevers and colds. It is also very useful in the treatment of restlessness and nervousness, being very useful as a mild nervine for children. A tea made from the leaves can also be used. The infusion is also applied externally to bruises, especially black eyes (Zargari, 2014; Mozaffarian, 2011; pfaf.org; Bhat *et al.*, 2018).

Nepetalactone: $C_{10}H_{14}O_2$

FIGURE 3.8 *Nepeta cataria* (flower, leaves, and whole plant) and its main component: Nepetalactone.

Ocimum basilicum (Figure 3.9) is a perennial that grows to a size of 0.5 m by 0.3 m at a fast rate. It is frost tender. The flowers are hermaphrodite and are pollinated by bees. Preference for light (sandy) and medium (loamy) soils, and well-drained soil. It can grow in acidic, neutral, and basic (alkaline) soils. It cannot grow in the shade. It prefers moist soil.

The leaves and flowers of "basil," raw or cooked, can be used as a flavoring or as a spinach. They are used especially with tomato dishes, pasta sauces, beans, peppers, and eggplant. The leaves are normally used fresh but can also be dried for winter use. A very pleasant addition to salads, the leaves have a delightful scent of cloves. Use the leaves sparingly in cooking as the heat concentrates the flavor. A refreshing tea is made from the leaves. The seed can be eaten on its own or added to bread dough as a flavoring. When soaked in water, it becomes mucilaginous and can be made into a refreshing beverage called "sherbet tokhum" in the Mediterranean. An essential oil obtained from the plant is used as a food flavoring in mustards, sauces, vinegars, etc. Basil contains estragole ($C_{10}H_{12}O$), a potentially carcinogenic and mutagenic essential oil. Do not take during pregnancy or give basil oil to small infants/children (Zargari, 2014; Mozaffarian, 2011; pfaf.org; Khair-ul-Bariyah *et al.*, 2012).

FIGURE 3.9 *Ocimum basilicum* (leaves, flowers, and whole plant).

Origanum majorana (Figure 3.10) is a perennial that grows to a size of 0.6 m by 0.6 m. The species is hermaphrodite and is pollinated by bees. It is noted for attracting wildlife. Preference for light (sandy), medium (loamy), and heavy (clay) soils, and well-drained soil. It can grow in acidic, neutral, and basic (alkaline) soils, also in semi-shade (light woodland) or no shade. It prefers dry or moist soil.

"Sweet marjoram" is mainly used as a culinary herb, but is also medicinally valuable due to its stimulant and anti-spasmodic properties. It is a good general tonic, treating various disorders of the digestive and respiratory systems. It has a stronger effect on the nervous system than the related oregano (*O. vulgare*) and is also thought to lower sex drive. Because it can promote menstruation, it should not be used medicinally by pregnant women though small quantities used for culinary purposes are safe. The herb is antiseptic, anti-spasmodic, carminative, cholagogue, diaphoretic, diuretic, emmenagogue, expectorant, stimulant, stomachic, and mildly tonic. It is taken internally in the treatment of bronchial complaints, tension headaches, insomnia, anxiety, minor digestive upsets, and painful menstruation. It should not be prescribed for pregnant women. Externally, it is used to treat muscular pain, bronchial complaints, arthritis, sprains, and stiff joints. The plant is harvested as flowering begins and can be used fresh or dried. Marjoram is often used medicinally in the form of the essential oil, about 400 grams being obtained from 70 kilos of the fresh herb. The oil is used as an external application for sprains, bruises, etc. The essential oil is used in aromatherapy (Zargari, 2014; Mozaffarian, 2011; pfaf.org; Bina & Rahimi, 2017).

Sabinene (Sabinen, Sabinin): $C_{10}H_{16}$

FIGURE 3.10   *Origanum majorana* (flowers and whole plant) and a main component: Sabinene.

Plantago major (Figure 3.11) is a perennial that grows to a size of 0.1 m at a medium rate. It is not frost tender. The flowers are hermaphrodite and are pollinated by wind. The plant is self-fertile. It is noted for attracting wildlife. Preference for light (sandy), medium (loamy), and heavy (clay) soils, and well-drained soil. It can grow in acidic, neutral, and basic (alkaline) soils. It cannot grow in the shade. It prefers moist soil. The plant can tolerate maritime exposure.

"Common plantain" is a safe and effective treatment for bleeding: it quickly staunches blood flow and encourages the repair of damaged tissue. The leaves are astringent, demulcent, deobstruent, depurative, diuretic, expectorant, hemostatic, and refrigerant. Internally, they are used in the treatment of a wide range of complaints including diarrhea, gastritis, peptic ulcers, irritable bowel syndrome, hemorrhage, hemorrhoids, cystitis, bronchitis, catarrh, sinusitis, asthma, and hay fever. They are used externally in treating skin inflammations, malignant ulcers, cuts, stings, etc. The heated leaves are used as a wet dressing for wounds, swellings, etc. The root is a remedy for the bite of rattlesnakes; it is used in equal portions with *Marrubium vulgare*. The seeds are used in the treatment of parasitic worms. Plantain seeds contain up to 30% mucilage which swells up in the gut, acting as a bulk laxative and soothing irritated membranes. Sometimes the seed husks are used without the seeds. A distilled water made from the plant makes an excellent eye lotion. High doses may cause a fall in blood pressure and diarrhea. Possible allergic contact dermatitis. Avoid in patients with intestinal obstruction or abdominal discomfort (Zargari, 2014; Mozaffarian, 2011; pfaf.org; Najafian *et al.*, 2018).

Plantago psyllium (Figure 3.12) is an annual that grows to a size of 0.6 m by 0.3 m. The species is hermaphrodite and is pollinated by wind. The plant is self-fertile. Preference for light (sandy), medium (loamy), and heavy (clay) soils, and well-drained

Aucubin: $C_{15}H_{22}O_9$

FIGURE 3.11 *Plantago major* (whole plant, flower spikes, and leaves) and its main component: Aucubin.

FIGURE 3.12   *Plantago psyllium* (flowers and seeds).

soil. It can grow in acidic, neutral, and basic (alkaline) soils. It cannot grow in the shade. It prefers dry or moist soil.

"Psyllium" has been used as a safe and effective laxative for thousands of years in Western herbal medicine. Both the dried seeds and seed husks are demulcent, emollient, and laxative. The seeds have a mucilaginous coat and swell to several times their volume when in water. The seeds and husks contain high levels of fiber; they expand and become highly gelatinous when soaked in water. By maintaining a high water content within the large bowel, they increase the bulk of the stool, easing its passage. They are used as a demulcent and as a bulk laxative in the treatment of constipation, dysentery, and other intestinal complaints, having a soothing and regulatory effect upon the system. Their regulatory effect on the digestive system means that they can also be used in the treatment of diarrhea and by helping to soften the stool, they reduce the irritation of hemorrhoids. The jelly-like mucilage produced when psyllium is soaked in water can absorb toxins within the large bowel. Thus, it helps to remove toxins from the body and can be used to reduce auto-toxicity. The macerated and decocted seeds yield a rich mucilage that is used in relieving skin irritations and reddened eyelids (Zargari, 2014; Mozaffarian, 2011; pfaf.org; Uribe *et al.*, 1985).

Plumbago europaea (Figure 3.13) is a perennial that grows up to a size of 1 m. The flowers are hermaphrodite. Preference for light (sandy) soils and well-drained soil. It can grow in acidic, neutral, and basic (alkaline) soils. It cannot grow in the shade. It prefers dry or moist soil.

The whole of "common leadwort," but especially the root, is acrid, emetic, odontalgic, sialagogue, and vesicant. Chewing the root produces copious salivation and is said to be of benefit in treating toothache. In addition, it is useful for treating obesity (Zargari, 2014; Mozaffarian, 2011; pfaf.org; Jaradat *et al.*, 2016).

Rubia tinctorum (Figure 3.14) is an evergreen perennial that grows to a size of 1 m by 1 m at a medium rate. The species is hermaphrodite. Preference for light (sandy) and medium (loamy) soils, and well-drained soil. It can grow in acidic, neutral, and basic (alkaline) soils, also semi-shade (light woodland) or no shade. It prefers dry or moist soil.

"Rose madder" (common madder) can tolerate maritime exposure. The root is aperient, astringent, cholagogue, diuretic, and emmenagogue. It is taken internally in the

Plumbagin: $C_{11}H_8O_3$

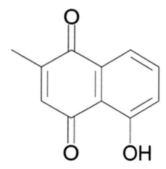

FIGURE 3.13 *Plumbago europaea* (flowers and leaves) and a main component: Plumbagin.

treatment of kidney and bladder stones. The root is seldom used in herbal medicine but is said to be effective in the treatment of amenorrhea, dropsy, and jaundice. The roots are harvested in the autumn from plants that are at least three years old. They are peeled and then dried. When taken internally the root imparts a red color to the milk, urine, and bones, especially the bones of young animals, and it is used in osteopathic investigations. It is used for fabric coloring in industry (Zargari, 2014; Mozaffarian, 2011; pfaf.org; Esalat Nejad & Esalat Nejad, 2013; Angelini *et al.*, 1997).

Sambucus ebulus (Figure 3.15) is a perennial that grows to a size of 1.2 m by 1 m at a fast rate. The species is hermaphrodite and is pollinated by bees, flies, and beetles. The plant is self-fertile. Preference for light (sandy), medium (loamy), and heavy (clay) soils, and can tolerate heavy clay soil. It can grow in acidic, neutral, and basic (alkaline) soils, also semi-shade (light woodland) or no shade. It prefers moist soil. The plant can tolerate strong winds but not maritime exposure. It can tolerate atmospheric pollution.

The leaves of "danewort" (dane weed) are anti-phlogistic, cholagogue, diaphoretic, diuretic, expectorant, and laxative. The fruit is also sometimes used, but it is less active than the leaves. The herb is commonly used in the treatment of liver and kidney complaints. When bruised and laid on boils and scalds, the leaves have a healing effect. The leaves can be made into a poultice for treating swellings and contusions.

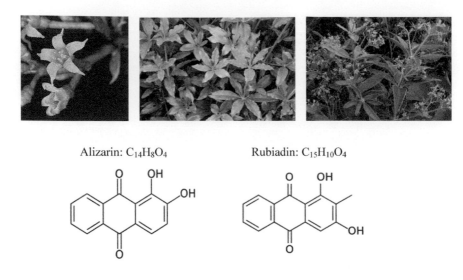

Alizarin: $C_{14}H_8O_4$ Rubiadin: $C_{15}H_{10}O_4$

Ruberythric acid: $C_{25}H_{26}O_{13}$ Purpurin (1,2,4-Trihydroxyanthraquinone): $C_{14}H_8O_5$

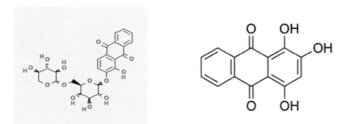

FIGURE 3.14 *Rubia tinctorum* (flowers, leaves, and whole plant) and its main components: Alizarin, rubiadin, ruberythric acid, purpurin.

FIGURE 3.15 *Sambucus ebulus* (whole plant, flowers, and fruits).

The leaves are harvested in the summer and can be dried for later use. The root is diaphoretic, mildly diuretic, and a drastic purgative. Dried, then powdered and made into a tea, it is one of the best remedies for dropsy. It should only be used with expert supervision as it can cause nausea and vertigo. A homeopathic remedy is made from the fresh berries or the bark. It is used in the treatment of dropsy (Zargari, 2014; Mozaffarian, 2011; pfaf.org; Shokrzadeh & Saeedi Saravi, 2010).

Sambucus nigra (Figure 3.16) is a deciduous bushy tree that grows to a size of 6 m by 6 m at a fast rate. The species is hermaphrodite and is pollinated by flies. Preference for light (sandy), medium (loamy), and heavy (clay) soils, and can tolerate heavy clay soil. It can grow in acidic, neutral, and basic (alkaline) soils, also very alkaline soils and semi-shade (light woodland) or no shade. It prefers moist soil. The plant can tolerate maritime exposure. It can tolerate atmospheric pollution.

"Elder" has a very long history of household use as a medicinal herb and is also much used by herbalists. The plant has been called "the medicine chest of country people." The flowers are the main part used in modern herbalism, though all parts of the plant have been used at times. It is a stimulant. The inner bark is collected from young trees in the autumn and is best sun-dried. It is diuretic, a strong purgative, and in large doses emetic. It is used in the treatment of constipation and arthritic conditions. An emollient ointment is made from the green inner bark. The leaves can be used both fresh and dry. For drying, they are harvested in periods of fine weather during June and July. The leaves are purgative, but are more nauseous than the bark. They are also diaphoretic, diuretic, expectorant, and hemostatic. The juice is said to be a good treatment for inflamed eyes. An ointment made from the leaves is emollient and is used in the treatment of bruises, sprains, chilblains, wounds, etc. The fresh flowers are used in the distillation of "elder flower water." The flowers can be preserved with salt to make them available for distillation later in the season. The water is mildly astringent and a gentle stimulant. It is mainly used as a vehicle for eye and skin lotions. The dried flowers are diaphoretic, diuretic, expectorant, galactogogue, and pectoral. An infusion is very effective in the treatment of chest complaints and is also used to bathe inflamed eyes. The infusion is also a very good spring tonic and blood cleanser. Externally, the flowers are used in poultices to ease pain and abate

Sambunigrin: $C_{14}H_{17}NO_6$

FIGURE 3.16 *Sambucus nigra* (flowers, fruits, and shrub) and its main component: Sambunigrin.

inflammation. Used as an ointment, it treats chilblains, burns, wounds, scalds, etc. The fruit is depurative, weakly diaphoretic, and gently laxative. A tea made from the dried berries is said to be a good remedy for colic and diarrhea. The fruit is widely used for making wines, preserves, etc., and these are said to retain the medicinal properties of the fruit. The pith of young stems is used in treating burns and scalds (Zargari, 2014; Mozaffarian, 2011; pfaf.org; Thi ho, 2017).

Sambucus racemosa (Figure 3.17) is a deciduous bushy tree that grows to a size of 4 m. The species is hermaphrodite and is pollinated by insects. Preference for light (sandy), medium (loamy), and heavy (clay) soils, and can tolerate heavy clay soil. It can grow in acidic, neutral, and basic (alkaline) soils, also semi-shade (light woodland) or no shade. It prefers moist soil. The plant can tolerate strong winds but not maritime exposure. It can tolerate atmospheric pollution.

The bark and the leaves of the "red elderberry" are used as a diuretic and purgative. The blossoms have been used in the treatment of measles (Zargari, 2014; Mozaffarian, 2011; pfaf.org; Rojo *et al.*, 2003).

Satureja hortensis (Figure 3.18) is an annual that grows to a size of 0.4 m. The flowers are hermaphrodite and are pollinated by insects. The plant is self-fertile. It is noted for attracting wildlife. Preference for light (sandy) and medium (loamy) soils, and well-drained soil. It can grow in acidic, neutral, and basic (alkaline) soils, and can tolerate very alkaline soils. It cannot grow in the shade. It prefers dry or moist soil and can tolerate drought.

"Summer savory" is most often used as a culinary herb, but it also has marked medicinal benefits, especially upon the whole digestive system. The plant has a milder action than the closely related "winter savory," *S. montana*. The whole herb, and especially the flowering shoots, is antiseptic, aromatic, carminative, digestive, expectorant, and stomachic. Taken internally, it is said to be a sovereign remedy for colic and a cure for flatulence, whilst it is also used to treat nausea, diarrhea, bronchial congestion, sore throats, and menstrual disorders. It should not be prescribed for pregnant women. A sprig of the plant, rubbed onto bee or wasp stings, brings instant relief. The plant is harvested in the summer when in flower and can be used fresh or dried. The essential oil forms an ingredient in

Sucrose: $C_{12}H_{22}O_{11}$

FIGURE 3.17 *Sambucus racemosa* (flowers, fruits, and shrub) and its main component: Sucrose.

Ursolic acid: $C_{30}H_{48}O_3$

FIGURE 3.18 *Satureja hortensis* (whole plant, flowers, and leaves) and its main component: Ursolic acid.

lotions for the scalp in cases of incipient baldness. An ointment made from the plant is used externally to relieve arthritic joints (Zargari, 2014; Mozaffarian, 2011; pfaf.org; Boroja *et al.*, 2018).

Valeriana officinalis (Figure 3.19) is a perennial that grows to a size of 1.5 m by 1 m. It is not frost tender. The flowers are hermaphrodite and are pollinated by bees, flies, and beetles. Preference for light (sandy), medium (loamy), and heavy (clay) soils. It can grow in acidic, neutral, and basic (alkaline) soils. It cannot grow in the shade. It prefers moist soil.

"Valerian" is a well-known and frequently used medicinal herb that has a long and proven history of efficacy. It is noted especially for its effect as a tranquillizer and nervine, particularly for those people suffering from nervous overstrain. Valerian has been shown to encourage sleep, improve sleep quality, and reduce blood pressure. It is also used internally in the treatment of painful menstruation, cramps, hypertension, irritable bowel syndrome, etc. It should not be prescribed for patients with liver problems. Externally, it is used to treat eczema, ulcers, and minor injuries. The root is anti-spasmodic, carminative, diuretic, hypnotic, powerfully nervine, sedative, and stimulant. The active ingredients are called valepotriates: research has confirmed that these have a calming effect on agitated people, but are also a stimulant in cases of fatigue. The roots of two-year-old plants are harvested in the autumn once the leaves have died down and are used fresh or dried. The fresh root is about three times as effective as roots dried at 40° (the report does not specify if this is centigrade or Fahrenheit), whilst temperatures above 82° destroy the active principle in the root. Use with caution; see the notes above on toxicity. It is said that prolonged medicinal use of this plant can lead to addiction. A course of treatment should not exceed three months. Adverse effects can include: Headaches (rare), giddiness, nausea,

Valerenic acid: $C_{15}H_{22}O_2$

FIGURE 3.19 *Valeriana officinalis* (whole plant, flowers, and leaves) and its main component: Valerenic acid.

excitability and agitation, heart palpitations (rare), insomnia (rare). Do not take with other sedatives (e.g. alcohol) or before driving (or alertness required) (Zargari, 2014; Mozaffarian, 2011; pfaf.org; Nandhini *et al.*, 2018).

Viburnum lantana (Figure 3.20) is a deciduous bushy tree that grows to a size of 5 m by 4 m at a medium rate. The species is hermaphrodite and is pollinated by insects. The plant is self-fertile. Preference for light (sandy), medium (loamy), and heavy (clay) soils, also well-drained soil and can tolerate heavy clay soil. It can grow in acidic, neutral, and basic (alkaline) soils and very alkaline soils, also semi-shade (light woodland) or no shade. It prefers dry or moist soil. It cannot tolerate atmospheric pollution.

"Wayfarer" contains triterpenoids, diterpenoids, and polyphenols, and is useful for treating diarrhea, sore throats, and gum inflammation (Zargari, 2014; Mozaffarian, 2011; pfaf.org; Sever Yilmaz *et al.*, 2007; Yilmaz *et al.*, 2008).

Viburnum opulus (Figure 3.21) is a deciduous bushy tree that grows to a size of 5 m by 5 m at a medium rate. The species is hermaphrodite and is pollinated by

FIGURE 3.20 *Viburnum lantana* (shrub, fruits, and flowers).

Scopoletin: $C_{10}H_8O_4$ Viburnine: $C_{27}H_{33}N_3O_3$

FIGURE 3.21   *Viburnum opulus* (shrub, flowers, and fruits) and its main components: Scopoletin, viburnine, and vitamin C (not shown).

insects. The plant is self-fertile. Preference for light (sandy), medium (loamy), and heavy (clay) soils, and can tolerate heavy clay soil. It can grow in acidic, neutral, and basic (alkaline) soils, and very alkaline soils, also in semi-shade (light woodland) or no shade. It prefers moist or wet soil.

"Guelder rose" is a powerful anti-spasmodic and is much used in the treatment of asthma, cramps, and other conditions such as colic or painful menstruation. It is also used as a sedative remedy for nervous conditions. The bark is anti-spasmodic, astringent, and sedative. The bark contains viburnine ($C_{27}H_{33}N_3O_3$), an alkaloid, as well as scopoletin ($C_{10}H_8O_4$), a coumarin that has a sedative effect on the uterus. A tea is used internally to relieve all types of spasms, including menstrual cramps, spasms after childbirth, and threatened miscarriage. It is also used in the treatment of nervous complaints and debility. The bark is harvested in the autumn before the leaves change color, or in the spring before the leaf buds open. It is dried for later use. The leaves and fruits are anti-scorbutic, emetic, and laxative. Berries are rich in vitamin C ($C_6H_8O_6$) and vitamin K. A homeopathic remedy is made from the fresh bark. It is used in the treatment of menstrual pain and spasms after childbirth (Zargari, 2014; Mozaffarian, 2011; pfaf.org; Yilmaz *et al.*, 2008).

4

Spermatophytes: Angiosperms, Dicotyledons, Monochlamydeaes

Spermatophytes are split into several sections. This chapter reviews angiosperms, dicotyledons, and monochlamydeaes, which are listed in Table 4.1. A total of 33 species is provided and described in detail below. As each species is presented, information on the taxonomy, plant use, and specifically the key natural products isolated from each plant is provided.

In this group, the flowers lack petals, but there are petals in several of the families. Another name for this group is apetales (Azadbakht, 2000).

Diversity of this group in Iran is moderate. It consists of three important families: cannabaceae, chenopodiaceae, and polygonaceae, all of which have important uses.

Alnus glutinosa (Figure 4.1) is a deciduous tree that grows to a size of 25 m by 10 m at a fast rate. The species is monoecious and pollinated by wind. It can fix nitrogen. It is noted for attracting wildlife. It has a preference for medium (loamy) soil, and can tolerate heavy clay and nutritionally poor soils. It can grow in acidic, neutral, and basic (alkaline) soils, as well as in semi-shade (light woodland) or no shade. It prefers moist or wet soil. The plant can tolerate maritime exposure.

The bark of "black alder" is alterative, astringent, cathartic, febrifuge, and tonic. The fresh bark will cause vomiting, so use dried bark for all but emetic purposes. A decoction of the dried bark is used to bathe swellings and inflammations, especially of the mouth and throat. The powdered bark and leaves have been used as an internal astringent and tonic, whilst the bark has also been used as an internal and external hemostatic against hemorrhage. The dried bark of young twigs is used, or the inner bark of branches that are 2–3 years old. It is harvested in the spring and dried for later use. Boiling the inner bark in vinegar produces a useful wash to treat lice and a range of skin problems such as scabies and scabs. The liquid can also be used as a tooth wash. The leaves are astringent, galactogogue, and vermifuge. They are used to help reduce breast engorgement in nursing mothers. A decoction of the leaves is used in folk remedies for treating cancer of the breast, duodenum, esophagus, face, pylorus, pancreas, rectum, throat, tongue, and uterus. The leaves are harvested in the summer and used fresh (Zargari, 2014; Mozaffarian, 2011; pfaf.org; Altinyay *et al.*, 2016).

Anabasis aphylla (Figure 4.2) is a perennial that grows to a size of 0.3 m. The species is hermaphrodite. It has a preference for light (sandy), medium (loamy), heavy (clay), and well-drained soil, and can tolerate nutritionally poor soil. It can grow in acidic, neutral, and basic (alkaline) soils. It cannot grow in the shade. It prefers dry or moist soil and can tolerate drought.

In industry, it is used for making insecticide. The annual branches contain anabasine ($C_{10}H_{14}N_2$), a botanical insecticide. Also, the plant is used for stabilizing sand dunes (Zargari, 2014; Mozaffarian, 2011; pfaf.org; Shakeri *et al.*, 2014).

TABLE 4.1

Spermatophytes: Angiosperms, Dicotyledons, Monochlamydeaes

Family Name	Species Name	Major Features of Botany	Usable Parts	Some Important Substances	Edible, Some Industrial and Medicinal Benefits
Betulaceae	*Alnus glutinosa*	Tree—Grey Bark Trunk	Bark of Trunk—Leaf	Tannic Acid	Dialysis—Alterative—Astringent—Cathartic—Febrifuge—Tonic—Galactogogue—Vermifuge Anti-Rheumatism—Treating Cancer of the Breast, Duodenum, and Esophagus
Betulaceae	*Betula alba*	Tree—White and Flat Bark Trunk—Diamond Leaf	Bark of Trunk—Wood	Betuloside—Betulin—Pentosans	Dialysis—Diuretic—Astringent—Lithontripic—Salve—Sedative—Anti-Rheumatism
Betulaceae	*Betula pendula*	Tree—White and Flat Bark Trunk—Diamond Leaf	Leaf—Bark of Trunk—Tar of Trunk—Bud—Vernal Sap—Shoot	Betuloside—Betulin—Verbascose	Anti-Rheumatism, Anti-Cholesterolemic, and Anti-Inflammatory—Cholagogue—Diaphoretic—Diuretic—Dialysis

(Continued)

TABLE 4.1 (CONTINUED)

Spermatophytes: Angiosperms, Dicotyledons, Monochlamydeaes

Family Name	Species Name	Major Features of Botany	Usable Parts	Some Important Substances	Edible, Some Industrial and Medicinal Benefits
Cannabaceae	*Cannabis sativa*	Herbal—Small Fruit—Long Leaf—Lots of Blossom	Whole Plant, Especially Leaf—Resin— Seed	Cannabin— Tetrahydrocannabinol— Linalool— Cannabidiol— Cannabinol—Pinene— Humulene	Mind Irritating—Anodyne—Anthelmintic—Cholagogue—Diuretic—Emollient—Hypnotic—Hypotensive—Laxative—Narcotic—Ophthalmic—Sedative—Anti-Emetic, Anti-Inflammatory, Anti-Periodic, and Anti-Spasmodic—Treating Alcohol Withdrawal, Anthrax, Asthma, Blood Poisoning, Bronchitis, Burns, Catarrh, Convulsions, Coughs, Cystitis, Delirium, Depression, Diarrhea, Dysentery, Dysmenorrhea, Epilepsy, Fever, Gonorrhea, Gout, Insomnia, Jaundice, Lockjaw, Malaria, Mania, Menorrhagia, Migraine, Morphine Withdrawal, Neuralgia, Palsy, Rheumatism, Scalds, Snake Bite, Swellings, Tetanus, Toothache, Uteral Prolapse
Cannabaceae	*Humulus lupulus*	Climber—Rough Trichome	Female Fruiting Body—Fruit— Flower—Female Flowering Head	Tannins—Lupulone— Lupulin—Humulone	Soothing—Sedative—Anodyne—Febrifuge—Hypnotic, Nervine, Stomachic—Calming—Tonic—Diuretic—Treating Paroxysm, Boils, Bruises, Calculus, Cancer, Cramps, Coughs, Cystitis, Debility, Delirium, Diarrhea, Dyspepsia, Fever, Fits, Hysteria, Nerves, Neuralgia, Rheumatism, and Worms—Antiseptic and Anti-Spasmodic

(Continued)

TABLE 4.1 (CONTINUED)

Spermatophytes: Angiosperms, Dicotyledons, Monochlamydeaes

Family Name	Species Name	Major Features of Botany	Usable Parts	Some Important Substances	Edible, Some Industrial and Medicinal Benefits
Chenopodiaceae	*Anabasis aphylla*	Herbal—Fleshy Stem	Young Twig	Anabasine—Aphylline—Lupinine	Industrial (Insecticide Making)
Chenopodiaceae	*Salicornia herbacea (S. europaea)*	Herbal—Without Leaf in Appearance—Red to Green	Leaf—Stem	Sterols—Oxalic Acid—Choline	Industrial (Green Salt Making)—Diuretic—Softener—Anti-Scurvy
Chenopodiaceae	*Spinacia oleracea*	Herbal—Triangular Leaf	Leaf—Seed	Spinacine—Neoxanthin—Violaxanthin	Edible—Carminative—Laxative—Cooling—Treating Asthma, Constipation, Febrile Conditions, Rheumatism, Inflammation of the Lungs, Bowels, and Liver
Corylaceae	*Carpinus betulus*	Tree—Oval and Long Leaf	Leaf—Bark of Trunk	Tannins	Hemostatic—Tonic—Astringent—Anti-Fever—Treating Sore throat
Corylaceae	*Corylus avellana*	Bushy Tree—Flower Appears in Spring Sooner than Leaf	Leaf—Fruit—Bark of Trunk—Seed—Catkin	Linoleic Acid—Vitamin B_1 and E—Phytosterols	Edible—Coagulator—Astringent—Diaphoretic—Febrifuge—Nutritive—Odontalgic—Anti-Varicose
Elaeagnaceae	*Elaeagnus angustifolia*	Bushy Tree—Small Flower	Seed—Flower—Fruit	Amino Acids—Fatty Acids—Fiber—Flavonoids—Vitamins A, C, E, and K	Edible—Astringent—Anti-Fever—Treating Catarrh, Cancer, and Bronchial Affections
Euphorbiaceae	*Euphorbia helioscopia*	Herbal—Small Red Flower—Oval Leaf	Root—Leaf—Stem—Seed	Diterpenoids—Diterpenes—Tannins—Amino Acids—Esters	Aperient—Emetic—Febrifuge Vermifuge—Anthelmintic—Cathartic—Anti-Asthma and Anti-Cancer

(Continued)

TABLE 4.1 (CONTINUED)

Spermatophytes: Angiosperms, Dicotyledons, Monochlamydeaes

Family Name	Species Name	Major Features of Botany	Usable Parts	Some Important Substances	Edible, Some Industrial and Medicinal Benefits
Euphorbiaceae	*Ricinus communis*	Bushy Tree—Evergreen—Big Glossy Leaf	Seed—Leaf—Root	Stearin—Ricin—Ricinoleic Acid	Industrial (Hydraulic and Brake Oil, Plastic, Polish Making)—Laxative—Aperient—Anthelmintic—Cathartic—Emollient, Purgative—Discutient—Expectorant—Anti-Tussive
Fagaceae	*Fagus sylvatica*	Tree—Leaf with Silky Trichome	Bark of Trunk—Branch	Creosol—Phlorol—Guaiacol—Creosote	Disinfectant—Antacid—Expectorant—Odontalgic—Stimulating—Anti-Fever, Anti-Indigestion, Anti-Pyretic, Antiseptic, and Anti-Tussive
Fagaceae	*Quercus brantii (Q. persica)*	Tree—Aspheric Female Flower—Hanging Male Flower	Leaf—Bark—Fruit—Gall—Seed—Seed Cups	Tannins—Quercetin—Quercite—Gallotannic Acid—Digallic Acid	Industrial (Tanning Leather and Ink Making)—Edible—Diuretic—Astringent—Styptic—Hemostatic—Anti-Bacterial, Anti-Fungal, and Antiseptic—Treating Tuberculosis, Hemorrhage, and Rickets
Juglandaceae	*Juglans regia*	Tree—Hanging Male Flower—Vertical Female Flower	Leaf—Seed—Bark of Root—Bark of Trunk—Fruit	Juglone—Citric Acid—Gallic Acid—Digallic Acid	Edible—Alterative—Anthelmintic—Anodyne—Astringent—Astringent—Depurative—Treating Diabetes, Constipation, Chronic Coughs, Asthma, Diarrhea, Dyspepsia, and Tuberculosis—Anti-Worm and Anti-Inflammatory
Lauraceae	*Laurus nobilis*	Tree or Bushy Tree—Evergreen—Leather Leaf	Leaf—Fruit	Pectic Acid—Essences—Resin—Laurostearin	Edible—Irritating—Aromatic—Astringent—Carminative—Diaphoretic—Digestive—Diuretic—Emetic—Emmenagogue—Narcotic Parasiticide—Stimulant—Stomachic—Stimulant—Treating Bloating—Antiseptic *(Continued)*

TABLE 4.1 (CONTINUED)

Spermatophytes: Angiosperms, Dicotyledons, Monochlamydeas

Family Name	Species Name	Major Features of Botany	Usable Parts	Some Important Substances	Edible, Some Industrial and Medicinal Benefits
Loranthaceae	*Viscum album*	Bushy Tree—Evergreen—Fleshy Leaf	Whole Plant, Especially Leaf—Twig	Tyramine—Lectins—Viscumin	Decreases Blood Pressure—Cardiac—Cytostatic—Diuretic—Hypotensive—Narcotic, Nervine—Stimulant—Tonic—Vasodilator—Anti-Spasmodic and Anti-Cancer—Treating Arthritis, Rheumatism, and Chilblains
Moraceae	*Ficus carica*	Tree—Rough Leaf	Fruit—Leaf—Latex	Vitamin C—Fiber—Omega 3—Carotene	Edible—Aperient—Laxative—Treating Insects Stings, Warts, and Cancer
Moraceae	*Morus alba*	Tree—Light Green Leaf	Leaf—Fruit—Stem—Branch—Stem of Root	Tannins—Linalool—Cannabinoids	Edible—Laxative—Astringent—Diaphoretic—Purgative—Hypoglycemic—Odontalgic—Ophthalmic—Diuretic—Hypotensive—Pectoral—Treating Chest Pain, Colds, Influenza, Eye Infections, Urinary Incontinence, Dizziness, Tinnitus, Insomnia, and Nosebleeds—Anti-Bacterial, Anti-Rheumatic, and Anti-Spasmodic
Moraceae	*Morus nigra*	Tree—Dark Green Leaf	Fruit—Leaf Branch—Stem—Stem of Root	Gallic Acid—Ellagic Acid—Anthocyanins	Edible—Analgesic—Emollient—Sedative—Diuretic—Hypotensive—Pectoral Astringent—Diaphoretic—Hypoglycemic—Odontalgic—Ophthalmic—Anti-Worm, Anti-Bacterial, Anti-Tussive, and Anti-Rheumatic—Treating Sore Throat
Nyctaginaceae	*Mirabilis jalapa*	Herbal—Flowers Available in a Range of Color	Root—Flower—Leaf	Betaxanthins—Rotenoids—Fatty Acids	Aperient—Aphrodisiac—Diuretic—Purgative—Anti-Worm—Treating Wounds

(Continued)

TABLE 4.1 (CONTINUED)

Spermatophytes: Angiosperms, Dicotyledons, Monochlamydeaes

Family Name	Species Name	Major Features of Botany	Usable Parts	Some Important Substances	Edible, Some Industrial and Medicinal Benefits
Platanaceae	*Platanus orientalis*	Tree—White Trunk	Bark—Leaf	Tannins	Astringent—Vulnerary—Treating Skin Rash, Ophthalmia, Diarrhea, Dysentery, Hernias, Toothache, and Voice Hoarseness
Polygonaceae	*Polygonum aviculare*	Herbal—Small and Narrow Leaf—With White Stipule	Whole Plant	Mucilage—Oxalic Acid	Strong Astringent—Anthelmintic—Astringent—Cardiotonic—Cholagogue—Diuretic—Expectorant—Febrifuge—Hemostatic—Lithontripic—Vasoconstrictor—Vulnerary—Treating Dysentery, Hemorrhoids, Wounds, Bleeding, Piles and Diarrhea
Polygonaceae	*Polygonum bistorta*	Herbal—With Rhizome	Root—Leaf—Rhizome	Tannins—Oxalic Acid—Vitamin C	Tonic—Strong Astringent—Demulcent—Diuretic—Febrifuge—Laxative—Strongly Styptic—Treating Internal and External Bleeding, Diarrhea, Dysentery, Cholera, Catarrh, Cystitis, Irritable Bowel Syndrome, Peptic Ulcers, Ulcerative Colitis
Polygonaceae	*Rumex acetosa*	Herbal—Sharp-Tip Leaf—Small Flower	Leaf—Seed—Root	Oxalic Acid	Edible—Appetizer—Astringent—Diuretic—Laxative—Refrigerant—Hemostatic—Depurative—Stomachic—Treating Gingivitis
Salicaseae	*Populus alba*	Tree—Strong Root—Leaf with White Trichome	Leaf—Bark of Trunk—Twig	Populin—Aricine—Salicin	Anodyne—Inflammatory—Astringent—Depurative—Diuretic—Tonic—Anti-Fever, Anti-Inflammatory, and Antiseptic—Treating Sciatica, Rheumatism, Arthritis, Gout, Urinary Complaints, Digestive—Debility and Anorexia

(Continued)

TABLE 4.1 (CONTINUED)

Spermatophytes: Angiosperms, Dicotyledons, Monochlamydeas

Family Name	Species Name	Major Features of Botany	Usable Parts	Some Important Substances	Edible, Some Industrial and Medicinal Benefits
Salicaseae	*Populus nigra*	Tree—Glossy and Green Leaf—White Wood	Bark of Trunk—Bud	Chrysene—Populin—Salicin—Salicylates	Anodyne—Astringent—Diuretic—Tonic—Treating Sciatica, Rheumatism, Kidney, and Bladder Diseases—Anti-Inflammatory and Antiseptic
Salicaceae	*Salix alba*	Tree—Silver White Leaf—Surface of the Leaf with White Hairs	Bark of Trunk—Leaf	Helicin—Butyric Acid—Salicin—Salicylic Acid	Astringent—Diaphoretic—Diuretic—Febrifuge—Hypnotic—Tonic—Anodyne—Sedative—Treating Fevers, Joint Pain, Rheumatic Ailments, and Headaches—Anti-Inflammatory, Anti-Periodic, Antiseptic, and Anti-Paroxysm
Thymelaeaceae	*Daphne mezereum*	Bushy Tree—Very Beautiful and Red Flower	Bark of Stem—Bark of Root—Fruit	Mezerein—Daphnin—Daphnetoxin	Diuretic—Laxative—Cathartic—Diuretic—Emetic—Rubefacient—Stimulant—Vesicant—Treating Rheumatism and Indolent Ulcers—Anti-Leukemia
Ulmaceae	*Celtis australis*	Tree—Dark and Oval Leaf with Trichome	Leaf—Fruit	Flavonoids—Phenolic Acids	Curing Diarrhea, Amenorrhea, Intermenstrual Bleeding, Colic, and Epilepsy—Astringent—Lenitive—Stomachic
Ulmaceae	*Ulmus campestris (U. glabra)*	Tree—Rough Bark—Leaf with Trichome	Bark of Trunk	Tannins—Mucilage—Phytosterols—Phlobaphenes	Tonic—Sweaty—Astringent—Demulcent—Diuretic—Anti-Cough—Treating Diarrhea, Rheumatism, Wounds, Eczema, and Piles
Urticaceae	*Urtica dioica*	Herbal—Oval Leaf	Whole Plant Especially Leaf and Root	Tannins—Mucilage—Formic Acid	Diuretic—Coagulator—Digestive—Tonic—Astringent—Depurative—Galactogogue—Hemostatic—Hypoglycemic—Treating Hay Fever, Arthritis, Skin Complaints, Arthritic Pain, Gout, Sciatica, Neuralgia, Hemorrhoids, Hair Problems, and Anemia—Anti-Asthmatic and Anti-Dandruff

FIGURE 4.1 *Alnus glutinosa* (tree, flowers, and leaf).

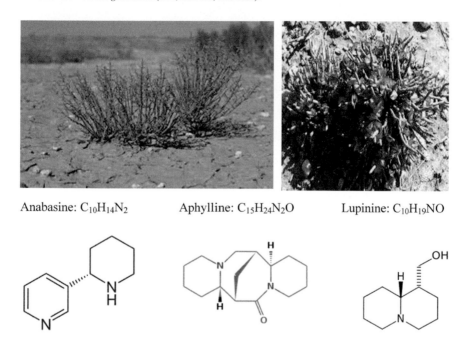

Anabasine: $C_{10}H_{14}N_2$ Aphylline: $C_{15}H_{24}N_2O$ Lupinine: $C_{10}H_{19}NO$

FIGURE 4.2 *Anabasis aphylla* (whole plant and flowers) and its main components: Anabasine, aphylline, lupinine.

Betula alba (Figure 4.3) is a deciduous tree that grows to a size of 20 m by 5 m at a fast rate. The species is monoecious and pollinated by wind. It has a preference for light (sandy), medium (loamy), and well-drained soil, and can tolerate heavy clay and nutritionally poor soils. It can grow in acidic, neutral, and basic (alkaline) soils. It cannot grow in the shade. It prefers dry or moist soil.

"Paper birch" was often employed medicinally by many native North American Indian tribes, who used it especially to treat skin problems. It is little used in modern herbalism. The bark is anti-rheumatic, astringent, lithontripic, salve, and sedative. The dried and powdered bark has been used to treat nappy rash in babies and various other skin rashes. A poultice of the thin outer bark has been used as a bandage on burns. A decoction of the inner bark has been used as a wash on rashes and other skin sores. Taken internally, the decoction has been used to treat dysentery and various diseases of the blood. The bark has been used to make casts for broken limbs.

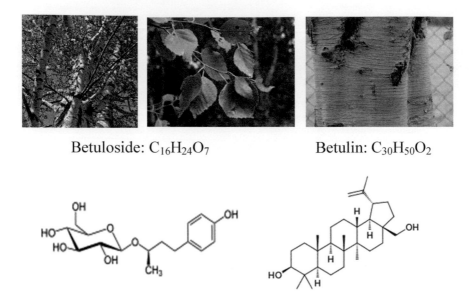

Betuloside: $C_{16}H_{24}O_7$ Betulin: $C_{30}H_{50}O_2$

FIGURE 4.3 *Betula alba* (tree, leaves, and trunk) and its main components: Betuloside, betulin.

A soft material such as a cloth is placed next to the skin over the broken bone. Birch bark is then tied over the cloth and is gently heated until it shrinks to fit the limb. A decoction of the wood has been used to induce sweating and to ensure an adequate supply of milk in a nursing mother. A decoction of both the wood and the bark has been used to treat female ailments (Zargari, 2014; Mozaffarian, 2011; pfaf.org; Vinod *et al.*, 2012).

Betula pendula (Figure 4.4) is a deciduous tree that grows to a size of 20 m by 10 m at a fast rate. The species is monoecious and pollinated by wind. It is noted for attracting wildlife. It has a preference for light (sandy), medium (loamy), and well-drained soil. It can tolerate heavy clay and nutritionally poor soils. It can grow in acidic, neutral, and basic (alkaline) soils and very acid soils. It cannot grow in the shade. It prefers dry or moist soil. The plant can tolerate strong winds but not maritime exposure.

"Silver birch" is anti-inflammatory, cholagogue, and diaphoretic. The bark is diuretic and laxative. An oil obtained from the inner bark is astringent and used in the treatment of various skin afflictions, especially eczema and psoriasis. The bark is usually obtained from trees that have been felled for timber and can be distilled at any time of the year. The inner bark is bitter and astringent and is used in treating intermittent fevers. The vernal sap is diuretic. The buds are balsamic. The young shoots and leaves secrete a resinous substance which has acid properties; when combined with alkalis it is a tonic laxative. The leaves are anti-cholesterolemic and diuretic; they also contain phytosides, which are effective germicides. An infusion of the leaves is used in the treatment of gout, dropsy, and rheumatism, and is recommended as a reliable solvent of kidney stones. The young leaves and leaf buds are harvested in the spring and dried for later use. A decoction of the leaves and bark is used for bathing skin eruptions. Moxa is made from the yellow fungous excrescences of the wood, which sometimes swell out of the fissures (Zargari, 2014; Mozaffarian, 2011; pfaf.org; Nagal *et al.*, 1983).

Verbascose: $C_{30}H_{52}O_{26}$

FIGURE 4.4 *Betula pendula* (tree, flowers, and leaf) and a main component: Verbascose.

Cannabis sativa (Figure 4.5) is an annual that grows to a size of 2.5 m by 0.8 m. The species is dioecious and pollinated by wind. The plant is not self-fertile. It is noted for attracting wildlife. It has a preference for light (sandy), medium (loamy), and heavy (clay) soils. It can grow in acidic, neutral, and basic (alkaline) soils, and can tolerate very acid and very alkaline soils. It cannot grow in the shade. It prefers moist soil.

"Hemp," or more appropriately "cannabis," since the form grown for fiber contains much less of the medicinally active compounds, has a very long history of medicinal use, though it is illegal to grow in many countries as the leaves and other parts of the plant are widely used as a narcotic drug. The leaves and the resin that exudes from them (cannabin: A greenish-black resin that is extracted from the dried leaves and flowering tops of the pistillate hemp plants and contains the physiologically active principles of cannabis), are the parts mainly used, though all parts of the plant contain the active ingredients. Cannabis contains a wide range of active ingredients, perhaps the most important of which is THC (tetrahydrocannabinol) ($C_{21}H_{30}O_2$). The principal uses of the plant are as a painkiller, sleep inducer, and reliever of the nausea caused by chemotherapy, whilst it also has a soothing influence in nervous disorders. Although cannabis does not effect a cure for many of the problems it is prescribed to treat, it is a very safe and effective medicine for helping to reduce the symptoms of many serious diseases. For example, it relieves the MS sufferer of the distressing desire to urinate, even when the bladder is empty. As long as it is used regularly, it also greatly reduces the pressure in the eye to relieve the symptoms of glaucoma. The whole plant is anodyne, anthelmintic, anti-emetic, anti-inflammatory, anti-periodic, anti-spasmodic, cholagogue, diuretic, emollient, hypnotic, hypotensive, laxative, narcotic, ophthalmic,

Tetra hydro cannabinol: $C_{21}H_{30}O_2$ Cannabidiol: $C_{21}H_{30}O_2$

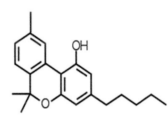

Cannabinol: $C_{21}H_{26}O_2$ Humulene: $C_{15}H_{24}$

FIGURE 4.5 *Cannabis sativa* (whole plant, leaves, and flower) and its main components: Tetrahydrocannabinol, cannabidiol, cannabinol, humulene.

and sedative. It is used to relieve some of the unpleasant side effects suffered by people undergoing chemotherapy for cancer—in particular it is very effective in removing the feelings of nausea and indeed helps to create an appetite and positive attitude of mind which is so important to people undergoing this treatment. It has also been found to be of use in the treatment of glaucoma and relieves the distressing constant desire to urinate that is suffered by many people with multiple sclerosis. Given to patients suffering from AIDS, it helps them to put on weight. Since it strongly increases the desire for food it has been found to be of benefit in treating anorexia nervosa. It is used externally as a poultice for corns, sores, varicose veins, gout, and rheumatism. Few plants have a greater array of folk medicine uses. Cannabis has been used in the treatment of a wide range of conditions including alcohol withdrawal, anthrax, asthma, blood poisoning, bronchitis, burns, catarrh, childbirth, convulsions, coughs, cystitis, delirium, depression, diarrhea, dysentery, dysmenorrhea, epilepsy, fever, gonorrhea, gout, inflammation, insomnia, jaundice, lockjaw, malaria, mania, menorrhagia, migraine, morphine withdrawal, neuralgia, palsy, rheumatism, scalds, snake bite, swellings, tetanus, toothache, uteral prolapse, and whooping cough. The seed is anodyne, anthelmintic, demulcent, diuretic, emollient, emmenagogue, febrifuge, laxative, narcotic, and tonic. It is used to treat constipation caused by debility or fluid retention. The seed is an important source of essential fatty acids and can be very helpful in the treatment of many nervous diseases. A high content of very active anti-bacterial and analgesic substances has been found in the plant. It has bactericidal effects on gram-positive micro-organisms, in some cases up to a dilution of 1:150,000 (Zargari, 2014; Mozaffarian, 2011; pfaf.org; El Sohly *et al.*, 2017).

Carpinus betulus (Figure 4.6) is a deciduous tree that grows to a size of 25 m by 20 m at a medium rate. The species is monoecious and pollinated by wind. It is noted for attracting wildlife. It has a preference for light (sandy) and medium (loamy) soil, and can tolerate heavy clay soil. It can grow in acidic, neutral, and basic (alkaline) soils, very alkaline soils, as well as in full shade (deep woodland), semi-shade (light woodland), or no shade. It prefers moist soil.

The leaves of the "European hornbeam" are hemostatic. They are used in external compresses to stop bleeding and heal wounds. A distilled water made from the leaves is an effective eye lotion. The leaves are harvested in August and dried for later use. The plant is used in Bach flower remedies—the keywords for prescribing it are "tiredness," "weariness," and "mental and physical exhaustion" (Zargari, 2014; Mozaffarian, 2011; pfaf.org; Jafari Foutami *et al.*, 2018).

Celtis australis (Figure 4.7) is a deciduous tree that grows to a size of 20 m by 10 m at a medium rate. The species is hermaphrodite and pollinated by bees. It has a preference for light (sandy), medium (loamy), and well-drained soil, and can tolerate nutritionally poor soil. It can grow in acidic, neutral, and basic (alkaline) soils. It cannot grow in the shade. It prefers dry or moist soil and can tolerate drought.

The leaves and fruits of the "European nettle tree" (Mediterranean hackberry, lote tree, honeyberry) are astringent, lenitive, and stomachic. The leaves are gathered in early summer and dried for later use. The fruit, particularly before it is fully ripe, is

FIGURE 4.6 *Carpinus betulus* (tree, leaves, and flowers).

FIGURE 4.7 *Celtis australis* (tree, fruits, and leaves).

considered to be more effective medicinally. A decoction of both leaves and fruit is used in the treatment of amenorrhea, heavy menstrual and intermenstrual bleeding, and colic. The decoction can also be used to astringe the mucous membranes in the treatment of diarrhea, dysentery, and peptic ulcers (Zargari, 2014; Mozaffarian, 2011; pfaf.org; Badoni *et al.*, 2011).

Corylus avellana (Figure 4.8) is a deciduous bushy tree that grows to a size of 6 m by 3 m at a medium rate. The species is monoecious and pollinated by wind. The plant is not self-fertile. It is noted for attracting wildlife. It has a preference for light (sandy), medium (loamy), and heavy (clay) soils. It can grow in acidic, neutral, and basic (alkaline) soils, in very acid and very alkaline soils, as well as in semi-shade (light woodland) or no shade. It prefers moist soil. The plant can tolerate strong winds but not maritime exposure.

The bark, leaves, catkins, and fruits of "common hazel" are sometimes used medicinally. They are astringent, diaphoretic, febrifuge, nutritive, and odontalgic. The seed is stomachic and tonic. The oil has a very gentle but constant and effective action in cases of infection with threadworm or pinworm in babies and young children (Zargari, 2014; Mozaffarian, 2011; pfaf.org; Shahidi *et al.*, 2007).

Daphne mezereum (Figure 4.9) is a deciduous bushy tree that grows to a size of 1.5 m by 1.5 m at a medium rate. The species is hermaphrodite and pollinated by bees, flies, moths, and butterflies. The plant is self-fertile. It is noted for attracting wildlife. It has a preference for medium (loamy) soil and can tolerate heavy clay soil. It can grow in acidic, neutral, and basic (alkaline) soils, and in semi-shade (light woodland). It prefers moist soil.

"Mezereum" has been used in the past for treating rheumatism and indolent ulcers, but because of its toxic nature it is no longer considered to be safe. The plant contains various toxic compounds, including daphnetoxin ($C_{27}H_{30}O_8$) and mezerein ($C_{38}H_{38}O_{10}$), and has anti-leukemia effects. The bark is cathartic, diuretic, emetic, rubefacient, stimulant, and vesicant. The root bark is the most active medically, but the stem bark is also used. It has been used in an ointment to induce discharge in indolent ulcers and also has a beneficial effect upon rheumatic joints. The bark is not usually taken internally and even when used externally this should be done with extreme caution and not applied if the skin is broken. The bark is harvested in the autumn and dried for later use. The fruits have sometimes been used as a purgative. A homeopathic remedy is made from the plant, which is used in the treatment of various skin complaints and inflammations (Zargari, 2014; Mozaffarian, 2011; pfaf.org; Kosheleva & Nikonov, 1968).

FIGURE 4.8 *Corylus avellana* (shrub, fruits, and leaves).

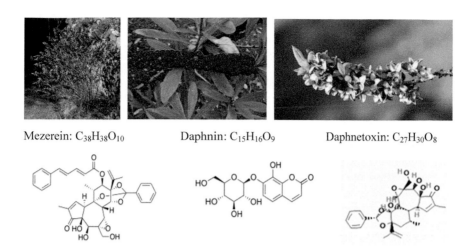

Mezerein: $C_{38}H_{38}O_{10}$ Daphnin: $C_{15}H_{16}O_9$ Daphnetoxin: $C_{27}H_{30}O_8$

FIGURE 4.9 *Daphne mezereum* (shrub, fruits, and flowers) and its main components: Mezerein, daphnin, daphnetoxin.

Elaeagnus angustifolia (Figure 4.10) is a deciduous bushy tree that grows to a size of 7 m by 7 m at a medium rate. The species is hermaphrodite and pollinated by bees. It can fix nitrogen. It has a preference for light (sandy), medium (loamy), heavy (clay), and well-drained soil, and can tolerate nutritionally poor soil. It can grow in acidic, neutral, and basic (alkaline) soils, and in very alkaline and saline soils. It cannot grow in the shade. It prefers dry or moist soil and can tolerate drought. The plant can tolerate maritime exposure.

The oil from the seeds of "Russian olive" is used with syrup as an electuary in the treatment of catarrh and bronchial affections. The juice of the flowers has been used in the treatment of malignant fevers. The fruit of many members of this genus is a very rich source of vitamins and minerals, especially vitamin A, vitamin C ($C_6H_8O_6$), vitamin E ($C_{29}H_{50}O_2$), flavonoids, and other bio-active compounds. It is also a fairly good source of essential fatty acids, which is unusual for a fruit. It is being investigated

FIGURE 4.10 *Elaeagnus angustifolia* (shrub and fruits).

as a food that can reduce the incidence of cancer and also as a means of halting or reversing the growth of cancers (Zargari, 2014; Mozaffarian, 2011; pfaf.org; Amiri Tehranizadeh *et al.*, 2016; Niknam et al., 2016).

Euphorbia helioscopia (Figure 4.11) is an annual that grows up to a size of 0.4 m. The species is hermaphrodite and pollinated by flies. It has a preference for light (sandy), medium (loamy), and well-drained soil. It can grow in acidic, neutral, and basic (alkaline) soils. It cannot grow in the shade. It prefers dry or moist soil. Anti-periodic.

The leaves and stems of the "sun spurge" are febrifuge and vermifuge. The root is anthelmintic. The plant is cathartic. It has anti-cancer properties. The milky sap is applied externally to skin eruptions. The seeds, mixed with roasted pepper, have been used in the treatment of cholera. The oil from the seeds has purgative properties (Zargari, 2014; Mozaffarian, 2011; pfaf.org; Saleem *et al.*, 2014).

Fagus sylvatica (Figure 4.12) is a deciduous tree that grows to a size of 30 m by 15 m at a medium rate. The species is monoecious and pollinated by wind. It is noted

FIGURE 4.11　*Euphorbia helioscopia* (whole plant, flowers, and leaves).

Creosol: $C_8H_{10}O_2$　　　Phlorol (2-Ethylphenol): $C_8H_{10}O$　　　Guaiacol: $C_7H_8O_2$

FIGURE 4.12　*Fagus sylvatica* (fruit and flowers) and its main components: Creosol, phlorol, guaiacol.

for attracting wildlife. It has a preference for light (sandy), medium (loamy), heavy (clay), and well-drained soil. It can grow in acidic, neutral, and basic (alkaline) soils, and can tolerate very acid and very alkaline soils. It can grow in full shade (deep woodland), semi-shade (light woodland), or no shade. It prefers dry or moist soil. The plant can tolerate strong winds but not maritime exposure. It can tolerate atmospheric pollution.

The bark of the "European beech" (common beech) is antacid, anti-pyretic, anti-septic, anti-tussive, expectorant, and odontalgic. A tar (or creosote), obtained by dry distillation of the branches, is stimulating and antiseptic. It is used internally as a stimulating expectorant and externally as an application to various skin diseases. The pure creosote has been used to give relief from toothache, but it should not be used without expert guidance. The plant is used in Bach flower remedies—the keywords for prescribing it are "intolerance," "criticism," and "passing judgements" (Zargari, 2014; Mozaffarian, 2011; pfaf.org; Petrakis *et al.*, 2011).

Ficus carica (Figure 4.13) is a deciduous tree that grows up to a size of 6 m at a medium rate. It is not frost tender. The flowers are monoecious. The plant is self-fertile. It has a preference for light (sandy), medium (loamy), and well-drained soil, and can tolerate heavy clay and nutritionally poor soils. It can grow in acidic, neutral, and basic (alkaline) soils. It cannot grow in the shade. It prefers dry or moist soil and can tolerate drought.

A decoction of the leaves of the "common fig" is stomachic. The leaves are also added to boiling water and used as a steam bath for painful or swollen piles. The latex from the stems is used to treat corns, warts, and piles. It also has an analgesic effect against insect stings and bites. The fruit is mildly laxative, demulcent, digestive, and pectoral. The unripe green fruits are cooked with other foods as a galactogogue and tonic. The roasted fruit is emollient and used as a poultice in the treatment of gum-boils, dental abscesses, etc. Syrup of figs, made from the fruit, is a well-known and effective gentle laxative that is also suitable for the young and very old. A decoction of the young branches is an excellent pectoral. The plant has anti-cancer properties. The sap and the half-ripe fruits are said to be poisonous. The sap can be a serious eye irritant (Zargari, 2014; Mozaffarian, 2011; pfaf.org; Mawa *et al.*, 2013).

Humulus lupulus (Figure 4.14) is a perennial climber that grows to a size of 6 m at a medium rate. The species is dioecious and pollinated by wind. The plant is not self-fertile. It is noted for attracting wildlife. It has a preference for light (sandy), medium (loamy), and heavy (clay) soils. It can grow in acidic, neutral, and basic (alkaline) soils, as well as in semi-shade (light woodland) or no shade. It prefers dry or moist soil and can tolerate drought.

FIGURE 4.13 *Ficus carica* (leaves and fruits).

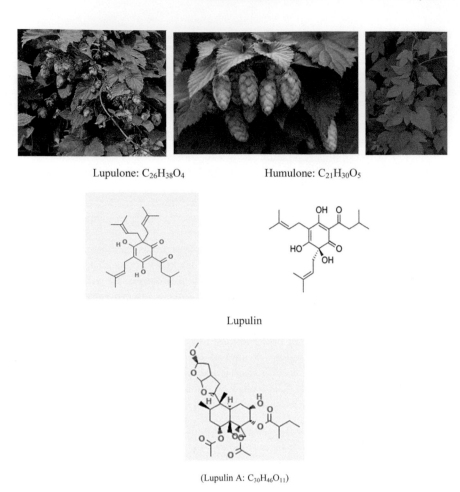

Lupulone: $C_{26}H_{38}O_4$ Humulone: $C_{21}H_{30}O_5$

Lupulin

(Lupulin A: $C_{30}H_{46}O_{11}$)

FIGURE 4.14 *Humulus lupulus* (whole plant, flowers, and leaves) and the its main components: Lupulone, humulone, lupulin.

The "common hop" has a long and proven history of herbal use, where it was employed mainly for its soothing, sedative, tonic, and calming effect on the body and mind. The strongly bitter flavor largely accounts for the ability to strengthen and stimulate the digestion, increasing gastric and other secretions. The female fruiting body is anodyne, antiseptic, anti-spasmodic, diuretic, febrifuge, hypnotic, nervine, sedative, stomachic, and tonic. Hops are widely used as a folk remedy to treat a wide range of complaints, including boils, bruises, calculus, cancer, cramps, cough, cystitis, debility, delirium, diarrhea, dyspepsia, fever, fits, hysteria, inflammation, insomnia, jaundice, nerves, neuralgia, rheumatism, and worms. The hairs on the fruits contain lupulin (lupulin A: $C_{30}H_{46}O_{11}$), a sedative and hypnotic drug. When given to nursing mothers, lupulin (lupulin A: $C_{30}H_{46}O_{11}$) increases the flow of milk—recent research has shown that it contains a related hormone that could account for this effect. The decoction from the flower is said to remedy swellings and hardness of the uterus. Hop flowers are much used as an infusion or can also be used to stuff pillows where the weight of the head releases the volatile oils. The fruit is also applied externally as a poultice to ulcers, boils, and painful swellings; it is said to remedy painful tumors. The female flowering heads are harvested in the autumn and can be used fresh or dried. Alcoholic extracts of hops in various dosage forms have been used clinically in treating numerous forms of leprosy, pulmonary tuberculosis, and acute bacterial dysentery, with varying degrees of success in China. The female fruiting body contains humulone ($C_{21}H_{30}O_5$) and lupulone ($C_{26}H_{38}O_4$), these are highly bacteriostatic against gram-positive and acid-fast bacteria. A cataplasm of the leaf is said to remedy cold tumors (Zargari, 2014; Mozaffarian, 2011; pfaf.org; Stevens *et al.*, 1997).

Juglans regia (Figure 4.15) is a deciduous tree that grows to a size of 20 m at a medium rate. It is not frost tender. The flowers are monoecious and pollinated by wind. The plant is self-fertile. It has a preference for light (sandy), medium (loamy), heavy (clay), and well-drained soil. It can grow in acidic, neutral, and basic (alkaline) soils. It cannot grow in the shade. It prefers moist soil.

The "walnut" tree has a long history of medicinal use, being used in folk medicine to treat a wide range of complaints. The leaves are alterative, anthelmintic, anti-inflammatory, astringent, and depurative. They are used internally the treatment of constipation, chronic coughs, asthma, diarrhea, dyspepsia, etc. The leaves are also used to treat skin ailments and purify the blood. They are considered to be specific in the treatment of strumous sores. Male inflorescences are made into a broth and used in the treatment of coughs and vertigo. The rind is anodyne and astringent. It is used in the treatment of diarrhea and anemia. The seeds are anti-lithic, diuretic, and stimulant. They are used internally in the treatment of lower back pain, frequent urination, weakness of both legs, chronic cough, asthma, constipation due to dryness or anemia,

Juglone: $C_{10}H_6O_3$ Digallic acid: $C_{14}H_{10}O_9$

FIGURE 4.15 *Juglans regia* (tree, fruit, and leaves) and its main components: Juglone, digallic acid.

and stones in the urinary tract. Externally, they are made into a paste and applied as a poultice to areas of dermatitis and eczema. The oil from the seed is anthelmintic. It is also used in the treatment of menstrual problems and dry skin conditions. The cotyledons are used in the treatment of cancer. The walnut has a long history of folk use in the treatment of cancer; some extracts from the plant have shown anti-cancer activity. The bark and rootbark are anthelmintic, astringent, and detergent. The plant is used in Bach flower remedies—the keywords for prescribing it are "oversensitive to ideas and influences" and the "link-breaker" (Zargari, 2014; Mozaffarian, 2011; pfaf .org; Panth *et al.*, 2016).

Laurus nobilis (Figure 4.16) is an evergreen tree or bushy tree that grows to a size of 12 m by 10 m at a slow rate. The species is dioecious and pollinated by bees. The plant is not self-fertile. It has a preference for light (sandy), medium (loamy), heavy (clay), and well-drained soil. It can grow in acidic, neutral, and basic (alkaline) soils, as well as in semi-shade (light woodland) or no shade. It prefers dry or moist soil. The plant can tolerate strong winds but not maritime exposure.

The "bay tree" has a long history of folk use in the treatment of many ailments, particularly as an aid to digestion and in the treatment of bronchitis and influenza. It has also been used to treat various types of cancer. The fruits and leaves are not usually administered internally, other than as a stimulant in veterinary practice, but were formerly employed in the treatment of hysteria, amenorrhea, flatulent colic, etc. Another report says that the leaves are used mainly to treat upper respiratory tract disorders and to ease arthritic aches and pains. It is settling to the stomach and has a tonic effect, stimulating the appetite and the secretion of digestive juices. The leaves are antiseptic, aromatic, astringent, carminative, diaphoretic, digestive, diuretic, emetic

Pectic acid (Polygalacturonic acid): $(C_6H_8O_6)_n$

FIGURE 4.16   *Laurus nobilis* (flowers and leaves) and its main component: Pectic acid.

in large doses, emmenagogue, narcotic, parasiticide, stimulant, and stomachic. The fruit is antiseptic, aromatic, digestive, narcotic, and stimulant. An infusion has been used to improve the appetite and as an emmenagogue. The fruit has also been used in making carminative medicines and was used in the past to promote abortion. A fixed oil from the fruit is used externally to treat sprains, bruises, etc., and is sometimes used as ear drops to relieve pain. The essential oil from the leaves has narcotic, antibacterial, and fungicidal properties (Zargari, 2014; Mozaffarian, 2011; pfaf.org; Dall Acqua *et al.*, 2009).

Mirabilis jalapa (Figure 4.17) is a perennial that grows to a size of 0.6 m by 0.5 m at a fast rate. The species is hermaphrodite and pollinated by insects. It has a preference for light (sandy), medium (loamy), heavy (clay), and well-drained soil. It can grow in acidic, neutral, and basic (alkaline) soils, as well as in semi-shade (light woodland) or no shade. It prefers moist soil.

The root of the "marvel of Peru" (four o'clock flower) is aphrodisiac, diuretic, and purgative. It is used in the treatment of dropsy. A paste of the root is applied as a poultice to treat scabies and muscular swellings. The juice of the root is used in the treatment of diarrhea, indigestion, and fevers. The powdered root, mixed with cornflour (*Zea mays*), is baked and used in the treatment of menstrual disorders. The leaves are diuretic, and used to reduce inflammation. A decoction of the leaves is used to treat abscesses. The leaf juice is used to treat wounds. Eight betaxanthins can be isolated from *M. jalapa* flowers and rotenoids can be isolated from the roots. Also, a fatty acid is found as a minor component in the seed oil (Zargari, 2014; Mozaffarian, 2011; pfaf.org; Wang *et al.*, 2002).

Morus alba (Figure 4.18) is a deciduous tree that grows to a size of 18 m by 10 m at a medium rate. The species is monoecious. The plant is self-fertile. It has a preference for light (sandy), medium (loamy), heavy (clay), and well-drained soil. It can grow in acidic, neutral, and basic (alkaline) soils, as well as in semi-shade (light woodland) or

FIGURE 4.17 *Mirabilis jalapa* (whole plant, flower, and fruit).

FIGURE 4.18 *Morus alba* (tree, fruits, and leaves).

no shade. It prefers moist soil and can tolerate drought. The plant can tolerate strong winds but not maritime exposure.

"White mulberry" has a long history of medicinal use in Chinese medicine, and almost all parts of the plant are used in one way or another. The leaves are anti-bacterial, astringent, diaphoretic, hypoglycemic, odontalgic, and ophthalmic. They are taken internally in the treatment of colds, influenza, eye infections, and nosebleeds. An injected extract of the leaves can be used in the treatment of elephantiasis and purulent fistulae. The leaves are collected after the first frosts of autumn and can be used fresh but are generally dried. The stems are anti-rheumatic, anti-spasmodic, diuretic, hypotensive, and pectoral. They are used in the treatment of rheumatic pains and spasms, especially of the upper half of the body, and high blood pressure. A tincture of the bark is used to relieve toothache. The branches are harvested in late spring or early summer and dried for later use. The fruit has a tonic effect on kidney energy. It is used in the treatment of urinary incontinence, dizziness, tinnitus, insomnia due to anemia, neurasthenia, hypertension, diabetes, premature greying of the hair, and constipation in the elderly. The root bark is anti-asthmatic, anti-tussive, diuretic, expectorant, hypotensive, and sedative. It is used internally in the treatment of asthma, coughs, bronchitis, edema, hypertension, and diabetes. The roots are harvested in the winter and dried for later use. The bark is anthelmintic and purgative, and is used to expel tapeworms. Extracts of the plant have anti-bacterial and fungicidal activity (Zargari, 2014; Mozaffarian, 2011; pfaf.org; Ercisli & Orhan, 2007).

Morus nigra (Figure 4.19) is a deciduous tree that grows to a size of 10 m by 15 m at a slow rate. The species is monoecious. The plant is self-fertile. It has a preference for light (sandy), medium (loamy), heavy (clay), and well-drained soil. It can grow in acidic, neutral, and basic (alkaline) soils, as well as semi-shade (light woodland) or no shade. It prefers moist soil. It can tolerate atmospheric pollution.

"Mulberry" has a long history of medicinal use in Chinese medicine, and almost all parts of the plant are used in one way or another. The white mulberry (*M. alba*) is normally used, but this species has the same properties. It is analgesic, emollient, and sedative. The leaves are anti-bacterial, astringent, diaphoretic, hypoglycemic, odontalgic, and ophthalmic. They are taken internally in the treatment of colds, influenza, eye infections, and nosebleeds. The leaves are collected after the first frosts of autumn and can be used fresh but are generally dried. The stems are anti-rheumatic, diuretic, hypotensive, and pectoral. A tincture of the bark is used to relieve toothache. The branches are harvested in late spring or early summer and dried for later use. The fruit

FIGURE 4.19 *Morus nigra* (fruits and leaves).

has a tonic effect on kidney energy. It is used in the treatment of urinary incontinence, tinnitus, premature greying of the hair, and constipation in the elderly. Its main use in herbal medicine is as a coloring and flavoring in other medicines. The root bark is anti-tussive, diuretic, expectorant, and hypotensive. It is used internally in the treatment of asthma, coughs, bronchitis, edema, hypertension, and diabetes. The roots are harvested in the winter and dried for later use. The bark is anthelmintic and purgative, and is used to expel tapeworms. Extracts of the plant have anti-bacterial and fungicidal activity. A homeopathic remedy is made from the leaves. It is used in the treatment of diabetes (Zargari, 2014; Mozaffarian, 2011; pfaf.org; Mazimba *et al.*, 2011).

Platanus orientalis (Figure 4.20) is a fast-growing deciduous tree reaching up to 30 m in size. The flowers are monoecious. It has a preference for light (sandy), medium (loamy), and heavy (clay) soils. It can grow in acidic, neutral, and basic (alkaline) soils. It cannot grow in the shade. It prefers moist soil and can tolerate drought. The plant can tolerate strong winds but not maritime exposure. It can tolerate atmospheric pollution.

The leaves of the "Old World sycamore" (Oriental plane) are astringent and vulnerary. The fresh leaves are bruised and applied to the eyes in the treatment of ophthalmia. A decoction is used to treat dysentery and a cream made from the leaves is used to heal wounds and chilblains. The leaves are harvested in the spring and summer and can be dried for later use. The bark is boiled in vinegar and then used in the treatment of diarrhea, dysentery, hernias, and toothache. In hot, dry climates, the hairs of the fruits and leaves are believed to cause an effect similar to hay fever (Zargari, 2014; Mozaffarian, 2011; pfaf.org; Thai *et al.*, 2016).

Polygonum aviculare (Figure 4.21) is an annual that grows to a size of 0.3 m. The species is hermaphrodite and pollinated by insects. The plant is self-fertile. It is noted for attracting wildlife. It has a preference for light (sandy), medium (loamy), and heavy (clay) soils. It can grow in acidic, neutral, and basic (alkaline) soils, and

FIGURE 4.20 *Platanus orientalis* (tree, leaves, and fruits).

FIGURE 4.21 *Polygonum aviculare* (whole plant, flower, and leaves).

can tolerate very acid soils. It can grow in semi-shade (light woodland) or no shade. It prefers moist soil. The plant can tolerate maritime exposure.

"Knotweed" is a safe and effective astringent and diuretic herb that is used mainly in the treatment of complaints such as dysentery and hemorrhoids. It is also taken in the treatment of pulmonary complaints because the silicic acid (H_4O_4Si) it contains strengthens connective tissue in the lungs. The whole plant is anthelmintic, astringent, cardiotonic, cholagogue, diuretic, febrifuge, hemostatic, lithontripic, and vulnerary. It was formerly widely used as an astringent both internally and externally in the treatment of wounds, bleeding, piles, and diarrhea. Its diuretic properties make it useful in removing stones. An alcohol-based preparation has been used with success to treat varicose veins of recent origin. The plant is harvested in the summer and early autumn and dried for later use. The leaves are anthelmintic, diuretic, and emollient. The whole plant is anthelmintic, anti-phlogistic, and diuretic. The juice of the plant is weakly diuretic, expectorant, and vasoconstrictor. Applied externally, it is an excellent remedy to stay bleeding of the nose and to treat sores. The seeds are emetic and purgative. Recent research has shown that the plant is a useful medicine for bacterial dysentery (Zargari, 2014; Mozaffarian, 2011; pfaf.org; Salama & Marraiki, 2009).

Polygonum bistorta (Figure 4.22) is a perennial that grows to a size of 0.5 m by 0.5 m at a fast rate. The species is hermaphrodite and pollinated by insects. It has a preference for light (sandy), medium (loamy), and heavy (clay) soils. It can grow in acidic, neutral, and basic (alkaline) soils, and can tolerate very acid soils. It can grow in semi-shade (light woodland) or no shade. It prefers moist or wet soil.

"Bistort" is one of the most strongly astringent of all herbs and it is used to contract tissues and staunch blood flow. The root is powerfully astringent, demulcent, diuretic, febrifuge, laxative, and strongly styptic. It is gathered in early spring when the leaves are just beginning to shoot, and then dried. It is much used, both internally and externally, in the treatment of internal and external bleeding, diarrhea, dysentery, cholera, etc. It is also taken internally in the treatment of a wide range of complaints including catarrh, cystitis, irritable bowel syndrome, peptic ulcers, ulcerative colitis, and excessive menstruation. Externally, it makes a good wash for small burns and wounds, and is used to treat pharyngitis, stomatitis, vaginal discharge, anal fissures, etc. A mouth wash or gargle is used to treat spongy gums, mouth ulcers, and sore throats. The leaves are astringent and have a great reputation in the treatment of wounds. In Chinese medicine, the rhizome is used for epilepsy, fever, tetanus, carbuncles, snake and mosquito bites, scrofula, and cramps in the hands and feet. Considered useful in diabetes (Zargari, 2014; Mozaffarian, 2011; pfaf.org; Dong *et al.*, 2014).

FIGURE 4.22 *Polygonum bistorta* (whole plant, flower, and leaves).

Populus alba (Figure 4.23) is a deciduous tree that grows to a size of 20 m by 12 m at a fast rate. The species is dioecious and pollinated by wind. The plant is not self-fertile. It is noted for attracting wildlife. It has a preference for light (sandy), medium (loamy), and well-drained soil, and can tolerate heavy clay soil. It can grow in acidic, neutral, and basic (alkaline) soils. It cannot grow in the shade. It prefers dry or moist soil. The plant can tolerate maritime exposure.

The stem bark of the "silver poplar" (white poplar) is anodyne, anti-inflammatory, antiseptic, astringent, diuretic, and tonic. The bark contains salicylates, from which the proprietary medicine aspirin is derived. It is used internally in the treatment of rheumatism, arthritis, gout, lower back pain, urinary complaints, digestive and liver disorders, debility, anorexia, and also to reduce fevers and relieve the pain of menstrual cramps. Externally, the bark is used to treat chilblains, hemorrhoids, infected wounds, and sprains. The bark is harvested from side branches or coppiced trees and dried for later use. The leaves are used in the treatment of caries of teeth and bones. The twigs are depurative (Zargari, 2014; Mozaffarian, 2011; pfaf.org; Haouat *et al.*, 2013).

Populus nigra (Figure 4.24) is a deciduous tree that grows to a size of 30 m by 20 m at a fast rate. The species is dioecious and pollinated by wind. The plant is not self-fertile. It is noted for attracting wildlife. It has a preference for light (sandy),

Populin (Populoside, Populine): $C_{20}H_{22}O_8$　　　　Aricine: $C_{22}H_{26}N_2O_4$

Salicin: $C_{13}H_{18}O_7$

FIGURE 4.23 *Populus alba* (tree, leaves, and flowers) and its main components: Populin, aricine, salicin.

Chrysene: $C_{18}H_{12}$

FIGURE 4.24   *Populus nigra* (tree, leaves, and flowers) and a main component: Chrysene.

medium (loamy), and well-drained soil, and can tolerate heavy clay soil. It can grow in acidic, neutral, and basic (alkaline) soils. It cannot grow in the shade. It prefers moist soil. The plant can tolerate strong winds but not maritime exposure.

The leaf buds of the "black poplar" are covered with a resinous sap that has a strong turpentine odor and a bitter taste. They also contain salicin ($C_{13}H_{18}O_7$), a glycoside that probably decomposes into salicylic acid (aspirin) in the body. The buds are anti-scorbutic, antiseptic, balsamic, diaphoretic, diuretic, expectorant, febrifuge, salve, stimulant, tonic, and vulnerary. They are taken internally in the treatment of bronchitis and upper respiratory tract infections, and stomach and kidney disorders. They should not be prescribed to patients who are sensitive to aspirin. Externally, the buds are used to treat colds, sinusitis, arthritis, rheumatism, muscular pain, and dry skin conditions. They can be put in hot water and used as an inhalant to relieve congested nasal passages. The buds are harvested in the spring before they open and dried for later use. The stem bark is anodyne, anti-inflammatory, antiseptic, astringent, diuretic, and tonic. The bark contains salicylates, from which the proprietary medicine aspirin is derived. It is used internally in the treatment of rheumatism, arthritis, gout, lower back pain, urinary complaints, digestive and liver disorders, debility, anorexia, and also to reduce fevers and relieve the pain of menstrual cramps. Externally, the bark is used to treat chilblains, hemorrhoids, infected wounds, and sprains. The bark is harvested from side branches or coppiced trees and dried for later use (Zargari, 2014; Mozaffarian, 2011; pfaf.org; Benaida *et al.*, 2013).

Quercus brantii (Q. persica) (Figure 4.25) is a deciduous tree that grows to a size of 8 m by 8 m at a fast rate. The flowers are pollinated by wind. It has a preference for light (sandy), medium (loamy), heavy (clay), and well-drained soil. It can grow in acidic, neutral, and basic (alkaline) soils, and can tolerate very alkaline soils. It cannot grow in the shade. It prefers dry or moist soil and can tolerate drought. The plant can tolerate maritime exposure.

"Brant's oak," the same as other species of oak, is used in the traditional medicine of many cultures, being valued especially for its tannins. Various parts of the plant can be used; most frequently it is the leaves, bark, seeds, seed cups, or the galls

Quercite (5-Deoxyinositol): $C_6H_{12}O_5$

FIGURE 4.25 *Quercus brantii* (tree, seeds, and leaves) and a main component: Quercite.

that are produced as a result of insect damage. A decoction or infusion is astringent, anti-bacterial, anti-fungal, antiseptic, styptic, and hemostatic. It is taken internally to treat conditions such as acute diarrhea, dysentery, and hemorrhages. Externally, it is used as a mouthwash to treat toothache or gum problems and is applied topically as a wash on cuts, burns, various skin problems, hemorrhoids, and oral, genital, and anal mucosa inflammation. Extracts of the plant can be added to ointments and used for the healing of cuts. In industry, it is used for tanning leather and ink making (Zargari, 2014; Mozaffarian, 2011; pfaf.org; Mozaffarian, 1996; Azizi *et al.*, 2014).

Ricinus communis (Figure 4.26) is an evergreen bushy tree that grows to a size of 1.5 m by 1 m at a fast rate. The species is monoecious and pollinated by wind. The plant is self-fertile. It has a preference for light (sandy), medium (loamy), and well-drained soil, and can tolerate heavy clay soil. It can grow in acidic, neutral, and basic (alkaline) soils. It cannot grow in the shade. It prefers moist soil.

The oil from the seed of the "castor bean" (castor oil plant), is a very well-known laxative that has been widely used for over 2000 years. It is considered to be fast, safe, and gentle, prompting a bowel movement in 3–5 hours, and is recommended for both the very young and the aged. It is so effective that it is regularly used to clear the digestive tract in cases of poisoning. It should not be used in cases of chronic constipation, where it might deal with the symptoms but does not treat the cause. The flavor is somewhat unpleasant, however, and it can cause nausea in some people. The oil has a remarkable anti-dandruff effect. The oil is well-tolerated by the skin and so is sometimes used as a vehicle for medicinal and cosmetic preparations. Castor oil congeals to a gel mass when the alcoholic solution is distilled in the presence of sodium salts of higher fatty acids. This gel is useful in the treatment of non-inflammatory skin diseases and is a good protective in cases of occupational eczema and dermatitis. The seed is anthelmintic, cathartic, emollient, laxative, and purgative. It is rubbed on the temple to treat headaches and is also powdered and applied to abscesses and various

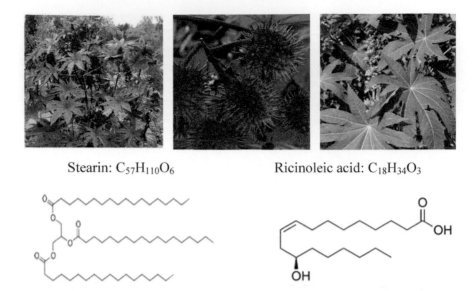

Stearin: $C_{57}H_{110}O_6$ Ricinoleic acid: $C_{18}H_{34}O_3$

FIGURE 4.26 *Ricinus communis* (shrub, flowers, and leaves) and its main components: Stearin, ricinoleic acid.

skin infections. The seed is used in Tibetan medicine, where it is considered to have an acrid, bitter, and sweet taste with a heating potency. It is used in the treatment of indigestion and as a purgative. A decoction of the leaves and roots is anti-tussive, discutient, and expectorant. The leaves are used as a poultice to relieve headaches and treat boils. In industry, it used for making hydraulic and brake oil (Zargari, 2014; Mozaffarian, 2011; pfaf.org; Jena & Gupta, 2012).

Rumex acetosa (Figure 4.27) is a perennial that grows to a size of 0.6 m by 0.3 m. It is not frost tender. The flowers are dioecious and pollinated by wind. The plant is not self-fertile. It is noted for attracting wildlife. It has a preference for light (sandy), medium (loamy), and heavy (clay) soils. It can grow in acidic, neutral, and basic (alkaline) soils, and can tolerate very acid soils. It can grow in semi-shade (light woodland) or no shade. It prefers moist soil.

The fresh or dried leaves of "sorrel" are astringent, diuretic, laxative, and refrigerant. They are used to make a cooling drink in the treatment of fevers and are especially useful in the treatment of scurvy. The leaf juice, mixed with fumitory, has been used as treatment for itchy skin and ringworm. An infusion of the root is astringent,

FIGURE 4.27 *Rumex acetosa* (whole plant, flowers, and leaves).

diuretic, and hemostatic. It has been used in the treatment of jaundice, gravel, and kidney stones. Both the roots and the seeds have been used to stem hemorrhages. A paste of the root is applied to set dislocated bones. The plant is depurative and stomachic. A homeopathic remedy is made from the plant, and is used in the treatment of spasms and skin ailments. Plants can contain quite high levels of oxalic acid $(C_2H_2O_4)$, which is what gives the leaves of many members of this genus an acidic lemon flavor. Perfectly alright in small quantities, the leaves should not be eaten in large amounts as the oxalic acid $(C_2H_2O_4)$ can lock up other nutrients in the food, especially calcium, thus causing mineral deficiencies. The oxalic acid $(C_2H_2O_4)$ content is reduced if the plant is cooked. People with a tendency for rheumatism, arthritis, gout, kidney stones, or hyperacidity should take especial caution if including this plant in their diet as it can aggravate their condition (Zargari, 2014; Mozaffarian, 2011; pfaf.org; Vasas *et al.*, 2015).

Salicornia herbacea (S. europaea) (Figure 4.28) is an annual that grows up to a size of 0.3 m. The species is hermaphrodite and pollinated by wind. It has a preference for light (sandy), medium (loamy), and heavy (clay) soils. It can grow in acidic, neutral, and basic (alkaline) soils, and can tolerate very alkaline and saline soils. It cannot grow in the shade. It prefers dry or moist soil. The plant can tolerate maritime exposure.

"Glasswort" contains choline $(C_5H_{14}NO)$, oxalic acid $(C_2H_2O_4)$, and sterols and is diuretic, softener, and anti-scurvy. In industry, the leaf and stem of the plant are used for making green salt (Zargari, 2014; Mozaffarian, 2011; pfaf.org; Mozaffarian, 1996; Rhee *et al.*, 2009).

Salix alba (Figure 4.29) is a deciduous tree that grows to a size of 25 m by 10 m at a fast rate. It is not frost tender. The flowers are dioecious and pollinated by bees. The plant is not self-fertile. It is noted for attracting wildlife. It has a preference for light (sandy), and medium (loamy) soil , and can tolerate heavy clay soil. It can grow in acidic and neutral soils. It cannot grow in the shade. It prefers moist or wet soil. The plant can tolerate maritime exposure.

"White willow," famous as the original source of salicylic acid $(C_7H_6O_3)$ (the precursor of aspirin), along with several closely related species, has been used for thousands of years to relieve joint pain and manage fevers. The bark is anodyne, anti-inflammatory, anti-periodic, antiseptic, astringent, diaphoretic, diuretic, febrifuge, hypnotic, sedative, and tonic. It has been used internally in the treatment of dyspepsia connected with debility of the digestive organs, rheumatism, arthritis, gout, inflammatory stages of auto-immune diseases, feverish illnesses, neuralgia, and headaches. Its tonic and astringent properties render it useful in convalescence from acute diseases, and in treating worms, chronic dysentery, and diarrhea. The fresh bark is very

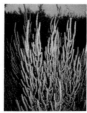

FIGURE 4.28 *Salicornia herbacea* (whole plant and stems).

Helicin: $C_{13}H_{16}O_7$

FIGURE 4.29   *Salix alba* (tree, leaves, and flower) and a main component: Helicon.

bitter and astringent; it contains salicin ($C_{13}H_{18}O_7$), which probably decomposes into salicylic acid ($C_7H_6O_3$) in the human body. This is used as an anodyne and febrifuge. The bark is harvested in the spring or early autumn from branches that are 3–6 years old and dried for later use. The leaves are used internally in the treatment of minor feverish illnesses and colic. An infusion of the leaves has a calming effect and is helpful in the treatment of nervous insomnia. When added to the bath water, the infusion is of real benefit in relieving widespread rheumatism. The leaves can be harvested throughout the growing season and are used fresh or dried. The German Commission E Monographs: Therapeutic guide to herbal medicines (Blumenthal *et al.*, 1998), approves Salix for diseases accompanied by fever, rheumatic ailments, and headaches. Gastrointestinal bleeding and kidney damage are possible. Avoid concurrent administration with other aspirin-like drugs. Avoid during pregnancy. Drug interactions associated with salicylates are applicable (Zargari, 2014; Mozaffarian, 2011; pfaf.org; Zarger *et al.*, 2014).

Spinacia oleracea (Figure 4.30) is a fast-growing annual reaching up to 0.3 m in size. The species is dioecious and pollinated by wind. The plant is not self-fertile. It has a preference for light (sandy), medium (loamy), and heavy (clay) soils. It can grow in acidic, neutral, and basic (alkaline) soils, as well as semi-shade (light woodland) or no shade. It prefers moist soil.

"Spinach" is carminative and laxative. In experiments, it has been shown to have hypoglycemic properties. It has been used in the treatment of urinary calculi. The leaves have been used in the treatment of febrile conditions, and inflammation of the lungs and bowels. The seeds are laxative and cooling, and have been used in the treatment of difficult breathing, inflammation of the liver, and jaundice (Zargari, 2014; Mozaffarian, 2011; pfaf.org; Kumar & Patwa, 2018).

Ulmus campestris (U. glabra) (Figure 4.31) is a deciduous tree that grows to a size of 30 m by 25 m at a fast rate. The species is hermaphrodite and pollinated by

Neoxanthin: $C_{40}H_{56}O_4$ Spinacine: $C_7H_9N_3O_2$

Violaxanthin: $C_{40}H_{56}O_4$

FIGURE 4.30 *Spinacia oleracea* (whole plant, leaves, and flowers) and its main components: Neoxanthin, spinacine, violaxanthin.

FIGURE 4.31 *Ulmus campestris* (tree, flower, and leaves).

wind. It is noted for attracting wildlife. It has a preference for light (sandy), medium (loamy), and well-drained soil, and can tolerate heavy clay soil. It can grow in acidic, neutral, and basic (alkaline) soils, as well as in semi-shade (light woodland) or no shade. It prefers moist soil. The plant can tolerate maritime exposure and atmospheric pollution.

The inner bark of the "wych elm" is astringent, demulcent, and mildly diuretic. It is used both internally and externally in the treatment of diarrhea, rheumatism, wounds, piles, etc., and is also used as a mouthwash in the treatment of ulcers. The inner bark is harvested from branches that are 3–4 years old and dried for later use.

The plant is used in Bach flower remedies—the keywords for prescribing it are "occasional feelings of inadequacy," "despondency," and "exhaustion from over-striving for perfection." A homeopathic remedy is made from the inner bark, and is used in the treatment of eczema (Zargari, 2014; Mozaffarian, 2011; pfaf.org; Mozaffarian, 1996; Boudaoud Ouahmed *et al.*, 2015).

Utrica dioica (Figure 4.32) is a perennial that grows to a size of 1.2 m by 1 m at a fast rate. The species is dioecious and pollinated by wind. The plant is not self-fertile. It is noted for attracting wildlife. It has a preference for light (sandy), medium (loamy), and heavy (clay) soils. It can grow in acidic, neutral, and basic (alkaline) soils, as well as in semi-shade (light woodland) or no shade. It prefers moist soil. The plant can tolerate strong winds but not maritime exposure.

The "common nettle" has a long history of use in the home as a herbal remedy and nutritious addition to the diet. A tea made from the leaves has traditionally been used as a cleansing tonic and blood purifier, so the plant is often used in the treatment of hay fever, arthritis, anemia, etc. The whole plant is anti-asthmatic, anti-dandruff, astringent, depurative, diuretic, galactogogue, hemostatic, hypoglycemic, and a stimulating tonic. An infusion of the plant is very valuable in stemming internal bleeding, and is also used to treat anemia, excessive menstruation, hemorrhoids, arthritis, rheumatism, and skin complaints, especially eczema. Externally, the plant is used to treat skin complaints, arthritic pain, gout, sciatica, neuralgia, hemorrhoids, hair problems, etc. The fresh leaves of nettles are rubbed or beaten onto the skin in the treatment of rheumatism, etc. This practice, called urtification, causes intense irritation to the skin as it is stung by the nettles. It is believed that this treatment works in two ways: Firstly, it acts as a counter-irritant, bringing more blood to the area to help remove the toxins that cause rheumatism. Secondly, the formic acid (CH_2O_2) from the nettles is believed to have a beneficial effect upon the rheumatic joints. For medicinal purposes, the plant is best harvested in May or June as it is coming into flower and is dried for later use. This species merits further study for possible uses against kidney and urinary system ailments. The juice of the nettle can be used as an antidote for stings from the leaves and an infusion of the fresh leaves is healing and soothing as a lotion for burns. The root has been shown to have a beneficial effect upon enlarged prostate glands. A homeopathic remedy is made from the leaves, and is used in the treatment of rheumatic gout, nettle rash, and chickenpox; externally is applied to bruises (Zargari, 2014; Mozaffarian, 2011; pfaf.org; Chandra Joshi *et al.*, 2014).

Viscum album (Figure 4.33) is an evergreen bushy tree that grows to a size of 1 m by 1 m. The species is dioecious. The plant is not self-fertile. It has a preference for light (sandy), medium (loamy), and heavy (clay) soils. It can grow in acidic, neutral, and basic (alkaline) soils, as well as in semi-shade (light woodland) or no shade. It prefers moist soil.

FIGURE 4.32 *Urtica dioica* (whole plant, flowers, and leaves).

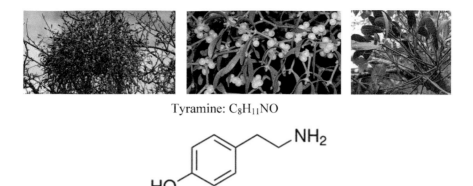

Tyramine: $C_8H_{11}NO$

FIGURE 4.33 *Viscum album* (shrub, fruits, and leaves) and its main component: Tyramine.

"Mistletoe" is chiefly used to lower blood pressure and heart rate, ease anxiety, and promote sleep. In low doses, it can also relieve panic attacks and headaches, and also improves the ability to concentrate. The plant's efficacy as an anti-cancer treatment has been subject to a significant amount of research—there is no doubt that certain constituents of the plant, especially the viscotoxins, exhibit anti-cancer activity but the value of the whole plant in cancer treatment is not fully accepted. It is said that the constituents of mistletoe vary according to the host plant it is growing on—those found on oak trees are said to be superior. Because of the potential side effects, this plant should only be used internally under the guidance of a skilled practitioner. Using the plant internally can provoke intolerant reactions to certain substances. The leaves and young twigs contain several medically active compounds. They are anti-spasmodic, cardiac, cytostatic, diuretic, hypotensive, narcotic, nervine, stimulant, tonic, and vasodilator. They are harvested just before the berries form and dried for later use. Mistletoe has a reputation for treating epilepsy and other convulsive nervous disorders. The effect of the correct dosage is to lessen and temporarily benumb the nervous activity that causes the spasms, but larger doses can produce the problem. Mistletoe has also been employed in checking internal hemorrhages, in treating high blood pressure, and in treating cancer of the stomach, lungs, and ovaries. Externally, the plant has been used to treat arthritis, rheumatism, chilblains, leg ulcers, and varicose veins. A homeopathic remedy is made from equal quantities of the berries and leaves; it is difficult to make because of the viscidity of the sap (Zargari, 2014; Mozaffarian, 2011; pfaf.org; Nazaruk & Orlikowski, 2016).

5

Spermatophytes: Angiosperms, Monocotyledons

Spermatophytes are split into several sections. This chapter reviews angiosperms and monocotyledons, which are listed in Table 5.1. A total of 23 species is provided and described in detail below. As each species is presented, information on the taxonomy, plant use, and specifically the key natural products isolated from each plant is provided.

In the embryo of these plants, there is a growing cotyledon. Their stems are not thick. The leaves of these plants, in most families, have parallel veins of leaf. The flower fragments are three or a multiple of three (Azadbakht, 2000).

Diversity of this group in Iran is low, but it consists of two important families: gramineae (poaceae) and liliaceae, both of which have important uses.

Allium akaka (Figure 5.1) is a bulb that grows to a size of 0.2 m by 0.1 m. The species is hermaphrodite and pollinated by bees and insects. It has a preference for light (sandy), medium (loamy), and well-drained soil. It can grow in acidic, neutral, and basic (alkaline) soils. It cannot grow in the shade. It prefers dry or moist soil.

Although no specific mention of medicinal uses has been seen for this species, members of this genus are in general very healthy additions to the diet. They contain sulfur compounds (which give them their onion flavor) and when added to the diet on a regular basis they help reduce blood cholesterol levels, act as a tonic to the digestive system, and also tonify the circulatory system (Zargari, 2014; Mozaffarian, 2011; pfaf .org; Sarmamy, 2018).

Allium cepa (Figure 5.2) is an evergreen bulb that grows up to a size of 0.6 m. It is not frost tender. The flowers are hermaphrodite and pollinated by bees and insects. It has a preference for light (sandy), medium (loamy), and well-drained soil. It can grow in acidic, neutral, and basic (alkaline) soils, and can tolerate very alkaline soils. It cannot grow in the shade. It prefers moist soil.

Although rarely used specifically as a medicinal herb, the "onion" has a wide range of beneficial actions on the body and when eaten (especially raw) on a regular basis will promote the general health of the body. The bulb is anthelmintic, anti-inflammatory, antiseptic, anti-spasmodic, carminative, diuretic, expectorant, febrifuge, hypoglycemic, hypotensive, lithontripic, stomachic, and tonic. When used regularly in the diet, it offsets tendencies towards angina, arteriosclerosis, and heart attack. It is also useful in preventing oral infection and tooth decay. Baked onions can be used as a poultice to remove pus from sores. Fresh onion juice is a very useful first aid treatment for bee and wasp stings, bites, grazes, or fungal skin complaints. When warmed, the juice can be dropped into the ear to treat earache. It also aids the formation of scar tissue on wounds, thus speeding up the healing process, and has been used as a cosmetic to remove freckles. Bulbs of red cultivars are harvested when mature in the summer

TABLE 5.1

Spermatophytes: Angiosperms, Monocotyledons

Family Name	Species Name	Major Features of Botany	Usable Parts	Some Important Substances	Edible, Some Industrial and Medicinal Benefits
Amaryllidaceae	*Narcissus tazetta*	Herbal—Aromatic Flower—Narrow and Long Leaf	Bulb	Essential Oils—Benzaldehyde	Emetic—Emmenagogue—Anti-Phlogistic—Treating Boils, Headaches, and Mastitis
Araceae	*Arum maculatum*	Herbal—With Rhizome—Red Fruit	Leaf—Root—Bark	Lycopene—Oxalic Acid	Aperient—Mucus Producing—Diaphoretic—Diuretic—Expectorant—Purgative—Vermifuge—Treating Diarrhea and Rheumatic Pain
Asparagaceae (recently, Liliaceae)	*Asparagus officinalis*	Herbal—Squamous Leaf—With Rhizome	Shoot—Root	Coniferin—Asparagusic Acid	Aperient—Cardiac—Demulcent—Diaphoretic—Diuretic—Sedative—Tonic—Treating Hepatitis—Anti-Spasmodic
Asparagaceae (recently, Liliaceae)	*Ruscus aculeatus*	Bushy Tree—Evergreen—With Rhizome—Small Flower in Green to White	Whole Plant—Root	Saponins—Sapogenins—Ruscogenin—Neoruscogenin	Appetizer—Diuretic—Kidney Stone Exertion—Aperient—Deobstruent—Depurative—Diaphoretic—Vasoconstrictor—Treating Gout, Hemorrhoids, and Jaundice—Anti-Inflammatory
Gramineae (Poaceae)	*Avena sativa*	Herbal—Long Fruit	Bud—Grain—Straw	Vitamins A and E—Beta Glucans—β-sitosterol	Edible—Tonic—Diuretic—Laxative—Wound Healing—Cardiac—Emollient, Nervine—Stimulant—Anti-Spasmodic, Antidepressant, and Anti-Tumor
Gramineae (Poaceae)	*Hordeum vulgare*	Herbal—Rough and Intermittent Leaf	Shoot—Bud—Grain—Stem—Bran—Straw	Albuminoids—Diastase—Amylose—Vitamins B_6, B_9, and B_{12}	Industrial (Paper, Chaff, Animal and Bird Food Making)—Edible—Tonic—Diuretic—Anti-Tumor and Anti-Fever—Treating Diarrhea and Nephritis

(Continued)

TABLE 5.1 (CONTINUED)

Spermatophytes: Angiosperms, Monocotyledons

Family Name	Species Name	Major Features of Botany	Usable Parts	Some Important Substances	Edible, Some Industrial and Medicinal Benefits
Gramineae (Poaceae)	*Oryza sativa*	Herbal—Large and Narrow Leaf	Rhizome—Grain	Starch—Vitamins B_1, B_2, and B_3	Industrial (Ethanol, Animal and Bird Food Making)—Edible—Nutritive—Soothing—Tonic—Diuretic—Lactation Reducer—Digestion Improver—Treating Diarrhea, Ulcers, and Night Sweats
Gramineae (Poaceae)	*Phragmites communis (Ph. australis)*	Herbal—With Rhizome—Big Leaf	Rhizome— Stem—Leaf	Fiber—Flavonoids	Industrial (Paper and Chaff Making)—Dialysis—Diuretic—Styptic—Refrigerant—Depurative—Febrifuge—Lithontripic—Sedative—Sialogogue Stomachic—Antidote, Anti-Emetic, Anti-Asthmatic, Anti-Tussive, and Anti-Pyretic—Treating Bronchitis and Cholera
Gramineae (Poaceae)	*Saccharum officinarum*	Herbal—Very Tall—Narrow and Tall Leaf	Stem—Leaf	Sacarose	Industrial (Animal and Bird Food, Paper, Ethanol, Sugar, and Enzymes Making)—Edible—Anti-Poison—Treating Sore Eyes, Sore Throats, Snake Bite, and Wounds
Gramineae (Poaceae)	*Secale cereale*	Herbal—Rough and Wide Leaf	Bud—Grain	Secalose—Starch—Gliadins	Edible—Laxative—Tonic—Softener—Treating Tumor, Blood Pressure, and Diabetes
Gramineae (Poaceae)	*Triticum sp.*	Herbal—Thin and Narrow Leaf	Bud—Grain—Stem	Glutens—Starch—Cellulose	Industrial (Paper, Chaff, Animal and Bird Food Making)—Edible—Tonic—Treating Asthma and Inflammation—Anti-Poison

(Continued)

TABLE 5.1 (CONTINUED)

Spermatophytes: Angiosperms, Monocotyledons

Family Name	Species Name	Major Features of Botany	Usable Parts	Some Important Substances	Edible, Some Industrial and Medicinal Benefits
Gramineae (Poaceae)	*Zea mays*	Herbal—Rough and Tall Leaf—Glossy and Round Fruit	Leaf—Root—Cob—Silk—Grain	Maltose—Cerebronic Acid—Starch—Glucose—Albuminoids—Allantoin	Industrial (Paper, Animal and Bird Food Making)—Edible—Cholagogue—Demulcent—Diuretic—Lithontripic—Stimulant—Vasodilator—Coagulator—Treating Nephritis, Strangury, Dysuria, Diabetes, Cystitis, Gonorrhea, Gout, Gravel, and Swell
Iridaceae	*Crocus sativus*	Herbal—Hard Bulb—Violet Flower	Stigma—Style	Crocin—Crocetin	Edible—Anodyne—Sedative—Aphrodisiac—Appetizer—Carminative—Diaphoretic—Emmenagogue—Expectorant—Sedative—Stimulant—Treating Toothache—Anti-Spasmodic
Iridaceae	*Iris florentina*	Herbal—With Rhizome—Large, Fragrant and White Flower	Rhizome—Leaf	Vitamin C—Iridine—Irigenin—Irisin	Diuretic—Aperient—Treating Whooping Cough, Catarrh, and Diarrhea—Anti-Worm
liliaceae	*Allium akaka*	Herbal—With Bulb	Bulb—Leaf	Vitamins A and C—Sulfuronyl	Edible—Dialysis—Digestive—Tonic for Digestive System
liliaceae	*Allium cepa*	Herbal—With Bulb—Cylinder and Dark Green Leaf	Bulb	Citric Acid—Allylpropyl Disulfide—Anthocyanins	Edible—Anthelmintic—Carminative—Diuretic—Expectorant—Febrifuge—Hypoglycemic—Hypotensive—Lithontripic—Stomachic—Tonic—Antiseptic, Anti-Worm, Anti-Inflammatory, and Anti-Spasmodic-Treating Diabetes

(Continued)

TABLE 5.1 (CONTINUED)

Spermatophytes: Angiosperms, Monocotyledons

Family Name	Species Name	Major Features of Botany	Usable Parts	Some Important Substances	Edible, Some Industrial and Medicinal Benefits
liliaceae	*Allium porrum*	Herbal—With Bulb—Strip and Wide Leaf	Bulb—Leaf	Hemolysins—Maltase	Edible—Anthelmintic—Cholagogue—Diaphoretic—Diuretic—Expectorant—Febrifuge—Stimulant—Stomachic—Tonic—Vasodilator—Treating Indigestion, Obesity, and Joint Pain—Anti-Asthmatic, Anti-Cholesterolemic, Antiseptic, and Anti-Spasmodic
liliaceae	*Allium sativum*	Herbal—With Bulb—White Flower—Narrow Leaf	Bulb—Leaf	Allyl—Allylpropyl Disulfide—Propyl—Acrolein	Edible—Disinfectant—Appetizer—Decreases Blood Pressure—Anthelmintic—Cholagogue—Diaphoretic—Diuretic—Expectorant—Febrifuge—Stimulant—Stings—Stomachic—Tonic—Vasodilator—Anti-Worm, Anti-Asthmatic, Anti-Cholesterolemic, Antiseptic, and Anti-Spasmodic
liliaceae	*Asphodelus tenuifolius (A. fistulosus)*	Herbal—Lots of Leaves—Hollow Leaf—Capsule Fruit	Whole Plant—Seed	β-sitosterol—Octacosanol—Palmitic Acid—Tetracosanoic Acid—Linoleic Acid	Diuretic—Treating Atherosclerosis
liliaceae	*Lilium candidum*	Herbal—With Bulb—Narrow and Tall Leaf—Big White Flower	Bulb—Flower—Pollen	Cillin	Astringent—Demulcent, Emmenagogue—Emollient—Expectorant—Treating Hypothermia, Epilepsy, and Swell

(Continued)

TABLE 5.1 (CONTINUED)

Spermatophytes: Angiosperms, Monocotyledons

Family Name	Species Name	Major Features of Botany	Usable Parts	Some Important Substances	Edible, Some Industrial and Medicinal Benefits
Musaceae	*Musa sapientum* (*M. balbisiana*)	Herbal—Evergreen—Very Big and Spread Leaf	Fruit	Starch—Sacarose	Edible—Tonic—Diuretic—Anti-Worm
Orchidaceae	*Orchis maculata* (*O. fuchsii, O. longibracteata, Dactylorhiza maculata, D. fuchsii, D. longibracteata*)	Herbal—With Bulb—Pink to Purple Flower	Bulb	Bassorin—Arabian Gum	Industrial (Ice cream Making)—Treating Catching Cold, Bladder Swell, and Bloody Diarrhea
Palmaceae	*Phoenix dactylifera*	Tree—Big Trunk—Big Leaf	Fruit	Mucilage—Glucose—Zeaxanthin—Tannins—Vitamin A	Edible—Treating Cough and Chest Pain—Anti-Inflammatory and Cancer

FIGURE 5.1 *Allium akaka* (flower and leaves).

Allylpropyl disulfide: $C_6H_{12}S_2$

FIGURE 5.2 *Allium cepa* (bulbs and flowers) and its main components: Allylpropyl disulfide, citric acid (not shown).

and used to make a homeopathic remedy. This is used particularly in the treatment of people whose symptoms include running eyes and nose. There have been cases of poisoning caused by the consumption of this plant in large quantities and by some mammals. Dogs seem to be particularly susceptible. Hand eczema may occur with frequent handling. May interfere with drug control of blood sugar (Zargari, 2014; Mozaffarian, 2011; pfaf.org; Arora *et al.*, 2017).

Allium porrum (Figure 5.3) is a bulb that grows up to a size of 0.9 m. It is not frost tender. The flowers are hermaphrodite and pollinated by bees and insects. It has a preference for light (sandy), medium (loamy), and well-drained soil, and can tolerate heavy clay soil. It can grow in acidic, neutral, and basic (alkaline) soils, as well as in very alkaline soils. It cannot grow in the shade. It prefers moist soil.

The "leek" has the same medicinal virtues as garlic, but in a much milder and less effective form. These virtues are as follows: Leek, the same as garlic, is used in a wide

FIGURE 5.3   *Allium porrum* (whole plant, bulbs, and flower).

range of ailments, particularly ailments such as ringworm, Candida, and vaginitis, where its fungicidal, antiseptic, tonic, and parasiticidal properties have proved of benefit. It is also said to have anti-cancer activity. Daily use of leek in the diet has been shown to have a very beneficial effect on the body, especially the blood system and the heart. For example, demographic studies suggest that garlic is responsible for the low incidence of arteriosclerosis in areas of Italy and Spain where consumption of the bulb is heavy. The bulb is said to be anthelmintic, anti-asthmatic, anti-cholesterolemic, antiseptic, anti-spasmodic, cholagogue, diaphoretic, diuretic, expectorant, febrifuge, stimulant, stomachic, tonic, and vasodilator. The crushed bulb may be applied as a poultice to ease the pain of bites, stings, etc. Although no individual reports regarding this species have been seen, there have been cases of poisoning caused by the consumption of certain members of this genus in large quantities and by some mammals (Zargari, 2014; Mozaffarian, 2011; pfaf.org; Fattorusso *et al.*, 2001).

Allium sativum (Figure 5.4) is a bulb that grows to a size of 0.6 m by 0.2 m. It is not frost tender. The flowers are hermaphrodite and pollinated by bees and insects. It has a preference for light (sandy), medium (loamy), and well-drained soil. It can grow in acidic, neutral, and basic (alkaline) soils, as well as in very alkaline soils. It cannot grow in the shade. It prefers dry or moist soil.

FIGURE 5.4   *Allium sativum* (whole plant, bulbs, and flower) and its main components: Allyl, propyl, acrolein, and allylpropyl disulfide (not shown).

"Garlic" has a very long folk history of use in a wide range of ailments, particularly ailments such as ringworm, Candida, and vaginitis, where its fungicidal, antiseptic, tonic, and parasiticidal properties have proved of benefit. The plant produces inhibitory effects on gram-negative germs of the typhoid-paratyphoid-enteritis group; indeed, it possesses outstanding germicidal properties and can keep amebic dysentery at bay. It is also said to have anti-cancer activity. It has also been shown that garlic aids detoxification of chronic lead poisoning. Daily use of garlic in the diet has been shown to have a very beneficial effect on the body, especially the blood system and the heart. For example, demographic studies suggest that garlic is responsible for the low incidence of arteriosclerosis in areas of Italy and Spain where consumption of the bulb is heavy. Recent research has also indicated that garlic reduces glucose metabolism in diabetics, slows the development of arteriosclerosis, and lowers the risk of further heart attacks in myocardial infarct patients. Externally, the expressed juice is an excellent antiseptic for treating wounds. The fresh bulb is much more effective medicinally than stored bulbs; extended storage greatly reduces the anti-bacterial action. The bulb is said to be anthelmintic, anti-asthmatic, anti-cholesterolemic, antiseptic, anti-spasmodic, cholagogue, diaphoretic, diuretic, expectorant, febrifuge, stimulant, stings, stomachic, tonic, and vasodilator. The German Commission E Monographs: Therapeutic guide to herbal medicines (Blumenthal *et al.*, 1998), approve *Allium sativum* for arteriosclerosis, hypertension, and high cholesterol levels. There have been cases of poisoning caused by the consumption of this species in large quantities and by some mammals. Dogs seem to be particularly susceptible. Avoid with anti-clotting medication. Breastfeeding may worsen a baby's colic. Avoid several weeks prior to surgery. Bad breath! (Zargari, 2014; Mozaffarian, 2011; pfaf.org; Majewski, 2014).

Arum maculatum (Figure 5.5) is a perennial that grows up to a size of 0.5 m. The species is monoecious and pollinated by flies. It has a preference for light (sandy), medium (loamy), and heavy (clay) soils. It can grow in acidic, neutral, and basic (alkaline) soils, as well as in full shade (deep woodland), semi-shade (light woodland), or no shade. It prefers moist soil.

"Cuckoo pint" has been little used in herbal medicine and is generally not recommended for internal use. The shape of the flowering spadix has a distinct sexual symbolism and the plant did have a reputation as an aphrodisiac, though there is no evidence to support this. The root is diaphoretic, diuretic, expectorant, strongly purgative, and vermifuge. It should be harvested in the autumn or before the leaves are produced in the spring. It can be stored fresh in a cellar in sand for up to a year or can be dried for later use. The plant should be used with caution; see earlier notes

FIGURE 5.5 *Arum maculatum* (whole plant, flower, and leaves).

on toxicity. The bruised fresh plant has been applied externally in the treatment of rheumatic pain. A liquid from the boiled bark has been used in the treatment of diarrhea. A homeopathic remedy is prepared from the root and leaves, and has been used in the treatment of sore throats (Zargari, 2014; Mozaffarian, 2011; pfaf.org; Colak *et al.*, 2009).

Asparagus officinalis (Figure 5.6) is a perennial that grows to a size of 1.5 m by 0.8 m. It is not frost tender. The flowers are dioecious and pollinated by bees. The plant is not self-fertile. It is noted for attracting wildlife. It has a preference for light (sandy), medium (loamy), heavy (clay), and well-drained soil. It can grow in acidic, neutral, and basic (alkaline) soils, as well as in very acidic, very alkaline, and saline soils. It can also grow in semi-shade (light woodland) or no shade. It prefers moist soil. The plant can tolerate maritime exposure.

"Asparagus" has been cultivated for over 2000 years as a vegetable and medicinal herb. Both the roots and the shoots can be used medicinally; they have a restorative and cleansing effect on the bowels, kidneys, and liver. The plant is anti-spasmodic, aperient, cardiac, demulcent, diaphoretic, diuretic, sedative, and tonic. The freshly expressed juice is used. The root is diaphoretic, strongly diuretic, and laxative. An infusion is used in the treatment of jaundice and congestive torpor of the liver. The strongly diuretic action of the roots made it useful in the treatment of a variety of urinary problems including cystitis. It is also used in the treatment of cancer. The roots are said to be able to lower blood pressure. The roots are harvested in late spring, after the shoots have been cut as a food crop, and are dried for later use. The seeds possess antibiotic activity. Another report says that the plant contains asparagusic acid ($C_4H_6O_2S_2$) which is nematocidal and is used in the treatment of schistosomiasis. Large quantities of the shoots can irritate the kidneys. The berries are mildly poisonous (Zargari, 2014; Mozaffarian, 2011; pfaf.org; Al Snaf, 2015).

Asphodelus tenuifolius (A. fistulosus) (Figure 5.7) is an annual/perennial that grows up to a size of 0.6 m. The species is hermaphrodite and pollinated by insects. It has a preference for light (sandy), medium (loamy), and well-drained soil, and can also tolerate nutritionally poor soil. It can grow in acidic, neutral, and basic (alkaline) soils, as well as semi-shade (light woodland) or no shade. It prefers dry or moist soil.

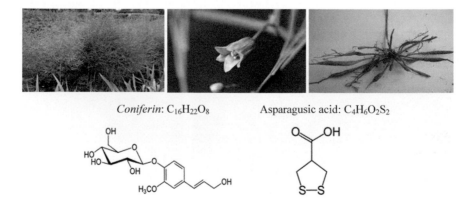

Coniferin: $C_{16}H_{22}O_8$ Asparagusic acid: $C_4H_6O_2S_2$

FIGURE 5.6 *Asparagus officinalis* (whole plant, flower, and root) and its main components: Coniferin, asparagusic acid.

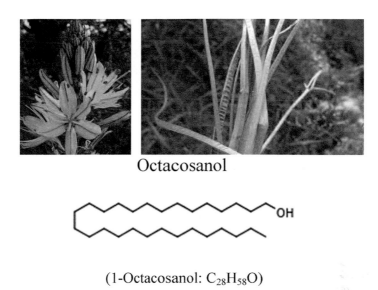

Octacosanol

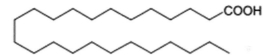

(1-Octacosanol: $C_{28}H_{58}O$)

Tetracosanoic acid (Lignoceric acid): $C_{24}H_{48}O_2$

FIGURE 5.7 *Asphodelus tenuifolius* (flowers and leaves) and its main components: Octacosanol, tetracosanoic acid.

The seed of "white asphodel" is diuretic. It is also applied externally to ulcers and inflamed parts of the body. The seed contains oils rich in linoleic acid ($C_{18}H_{32}O_2$) and are of value in preventing atherosclerosis (Zargari, 2014; Mozaffarian, 2011; pfaf.org; Mozaffarian, 1996; Safder *et al.*, 2009).

Avena sativa (Figure 5.8) is an annual that grows to a size of 0.9 m by 0.1 m. It is not frost tender. The flowers are hermaphrodite and pollinated by wind. The plant is self-fertile. It has a preference for light (sandy), medium (loamy), and well-drained

FIGURE 5.8 *Avena sativa* (whole plant, flowers, and grains).

soil, and can also tolerate heavy clay and nutritionally poor soils. It can grow in acidic, neutral, and basic (alkaline) soils, as well as very acid soils. It cannot grow in the shade. It prefers dry or moist soil and can tolerate drought.

"Oat" used mainly as a food; oat grain also has medicinal properties. In particular oats are a nutritious food that gently restores vigor after debilitating illnesses, helps lower cholesterol levels in the blood, and also increases stamina. The seed is a mealy nutritive herb that is anti-spasmodic, cardiac, diuretic, emollient, nervine, and stimulant. The seed contains the anti-tumor compound β-sitosterol ($C_{29}H_{50}O$) and has been used as a folk remedy for tumors. A gruel made from the ground seed is used as a mild nutritious aliment in inflammatory cases, fevers, and after parturition. It should be avoided in cases of dyspepsia accompanied with acidity of the stomach. A tincture of the ground seed in alcohol is useful as a nervine and uterine tonic. A decoction strained into a bath will help to soothe itchiness and eczema. A poultice made from the ground seeds is used in the treatment of eczema and dry skin. When consumed regularly, oat germ reduces blood cholesterol levels. Oat straw and the grain are prescribed to treat general debility and a wide range of nervous conditions. They are mildly antidepressant, gently raising energy levels and supporting an over-stressed nervous system. They are of particular value in helping a person to cope with the exhaustion that results from multiple sclerosis, chronic neurological pain, and insomnia. Oats are thought to stimulate sufficient nervous energy to help relieve insomnia. An alcoholic extraction of oats has been reported to be a deterrent for smoking, though reports that oat extract helped correct the tobacco habit have been disproven. A tincture of the plant has been used as a nerve stimulant and to treat opium addiction. The German Commission E Monographs: Therapeutic guide to herbal medicines (Blumenthal *et al.*, 1998), approve *Avena sativa* for inflammation of the skin, as well as warts (Zargari, 2014; Mozaffarian, 2011; pfaf.org; Das & Joseph, 2017).

Crocus sativus (Figure 5.9) is a corm that grows up to a size of 0.1 m. It is not frost tender. The flowers are hermaphrodite and pollinated by bees and butterflies. It has a preference for light (sandy), medium (loamy), and well-drained soil, and can also tolerate nutritionally poor soil. It can grow in acidic, neutral, and basic (alkaline) soils, as well as very alkaline soils. It can also grow in semi-shade (light woodland) or no shade. It prefers dry or moist soil.

"Saffron" is a famous medicinal herb with a long history of effective use, though it is little used at present because cheaper and more effective herbs are available. The flower styles and stigmas are the parts used, but as these are very small and fiddly to harvest they are very expensive and consequently often adulterated by lesser products. The styles and stigmas are anodyne, anti-spasmodic, aphrodisiac, appetizer, carminative, diaphoretic, emmenagogue, expectorant, sedative, and stimulant. They are used as a diaphoretic for children, to treat chronic hemorrhages in the uterus of adults, to induce menstruation, treat period pains, and calm indigestion and colic. A dental analgesic is obtained from the stigmas. The styles are harvested in the autumn when the plant is in flower and are dried for later use; they do not store well and should be used within 12 months. This remedy should be used with caution, as large doses can be narcotic and quantities of 10 g or more can cause an abortion. The plant is poisonous. The plant is perfectly safe in normal usage but 5–10 grams of saffron has been known to cause death (Zargari, 2014; Mozaffarian, 2011; pfaf.org; Ahmad Baba *et al.*, 2015).

Crocin: $C_{44}H_{64}O_{24}$

FIGURE 5.9 *Crocus sativus* (whole plant, flower, and bulbs) and its main components: Crocin, and crocetin (not shown).

Hordeum vulgare (Figure 5.10) is an annual that grows to a size of 1 m by 0.2 m. It is not frost tender. The flowers are hermaphrodite and pollinated by wind. It has a preference for light (sandy), medium (loamy), heavy (clay), and well-drained soil. It can grow in acidic, neutral, and basic (alkaline) soils. It cannot grow in the shade. It prefers moist soil. The plant can tolerate strong winds but not maritime exposure.

The shoots of "barley" are diuretic. The seed sprouts are demulcent, expectorant, galactofuge, lenitive, and stomachic. They are sometimes abortifacient. They are used in the treatment of dyspepsia caused by cereals, infantile lacto-dyspepsia, regurgitation of milk, and breast distension. They are best not given to a nursing mother as this can reduce milk flow. The seed is digestive, emollient, nutritive, febrifuge, and stomachic. It is taken internally as a nutritious food or as barley water (an infusion of the germinated seed in water) and is of special use for babies and people with physical disabilities. Its use is said to reduce excessive lactation. Barley is also used as a poultice for burns and wounds. The plant has a folk history of anti-tumor activity. The germinating seed has a hypoglycemic effect preceded by a hyperglycemic action. Modern research has shown that barley may be of aid in the treatment of hepatitis, whilst other trials have shown that it may help to control diabetes. Barley bran may have the effect of lowering blood cholesterol levels and preventing bowel cancer. Other uses are for bronchitis and diarrhea, and as a source of vitamin B_6, vitamin B_9

Amylose: $(C_6H_{10}O_5)_n - (H_2O)$

Vitamin B_{12} (Cobalamin): $C_{63}H_{88}CoN_{14}O_{14}P$

FIGURE 5.10 *Hordeum vulgare* (whole plant, spikelets, and grains) and its main components: Amylose, Vitamin B_{12}.

$(C_{19}H_{19}N_7O_6)$, and vitamin B_{12} $(C_{63}H_{88}CoN_{14}O_{14}P)$. Weight loss. Exposure to barley flour can cause asthma. Possible trigger for coeliac disease. Possible hypersensitivity to barley. In industry, it is used for making paper, chaff, animal and bird food (Zargari, 2014; Mozaffarian, 2011; pfaf.org; Duh *et al.*, 2001).

Iris florentina (Figure 5.11) is a perennial that grows to a size of 0.9 m by 0.6 m at a medium rate. It is not frost tender. The flowers are hermaphrodite and pollinated by insects. The plant is self-fertile. It has a preference for light (sandy) and medium (loamy) soils. It can grow in acidic, neutral, and basic (alkaline) soils, as well as in semi-shade (light woodland) or no shade. It prefers moist soil.

The dried root of "Florentine orris" (orris root), is diuretic, expectorant, and stomachic. It is taken internally in the treatment of coughs, catarrh, and diarrhea. Externally, it is applied to deep wounds. The root is harvested in late summer and early autumn and dried for later use. The juice of the fresh root is a strong purge of great efficiency in the treatment of dropsy. The leaves, and especially the rhizomes,

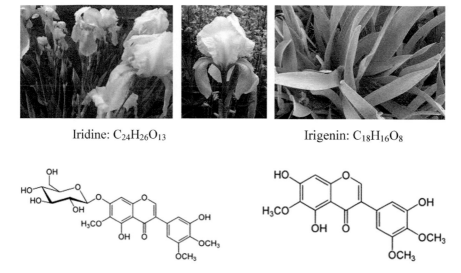

Iridine: $C_{24}H_{26}O_{13}$ Irigenin: $C_{18}H_{16}O_8$

FIGURE 5.11 *Iris florentina* (whole plant, flower, and leaves) and its main components: Iridine, irigenin, and vitamin C (not shown).

of this species contain an irritating resinous substance called irisin. If ingested, this can cause severe gastric disturbances. Plants can cause skin irritations and allergies in some people (Zargari, 2014; Mozaffarian, 2011; pfaf.org; Choudhary & Alam, 2017).

Lilium candidum (Figure 5.12) is a bulb that grows to a size of 1 m by 0.3 m. The species is hermaphrodite and pollinated by bees. It has a preference for light (sandy), medium (loamy), heavy (clay), and well-drained soil. It can grow in acidic, neutral, and basic (alkaline) soils, as well as in semi-shade (light woodland) or no shade. It prefers moist soil.

"Madonna lily" has a long history of herbal use, though it is seldom employed in modern herbalism because of its scarcity. The bulb and the flowers are astringent, highly demulcent, emmenagogue, emollient, and expectorant. The plant is mainly used externally, being applied as a poultice to tumors, ulcers, external inflammations, etc. The flowers are harvested when fully open and used fresh for making juice, ointments, or tinctures. The pollen has been used in the treatment of epilepsy (Zargari, 2014; Mozaffarian, 2011; pfaf.org; Munafo Jr & Gianfagna, 2015).

Musa sapientum (M. balbisiana) (Figure 5.13) is an evergreen perennial that grows to a size of 5 m by 3 m at a fast rate. The flowers are pollinated by birds and

FIGURE 5.12 *Lilium candidum* (whole plant, flower, and leaves).

Sacarose: $C_{12}H_{22}O_{11}$

$$CH_2OH \quad CH_2OH$$

FIGURE 5.13 *Musa sapientum* (whole plant, flowers, and fruits) and its main components: Sacarose, and starch (not shown).

bats. It is noted for attracting wildlife. It has a preference for light (sandy), medium (loamy), heavy (clay), and well-drained soil. It can grow in acidic, neutral, and very acidic soils, as well as in semi-shade (light woodland) or no shade. It prefers moist soil. The plant is not wind tolerant.

The fruit of the "banana" has medicinal properties. It has starch and sucrose. It is diuretic, tonic, and anti-worm (Zargari, 2014; Mozaffarian, 2011; pfaf.org; Mozaffarian, 1996; Imam & Akter, 2011).

Narcissus tazetta (Figure 5.14) is a bulb that grows to a size of 0.5 m by 0.1 m. It is not frost tender. The flowers are hermaphrodite and pollinated by bees. It has a preference for light (sandy) and medium (loamy) soil, and can tolerate heavy clay soil. It can grow in acidic, neutral, and basic (alkaline) soils, as well as in very alkaline soils. It cannot grow in the shade. It prefers moist soil.

FIGURE 5.14 *Narcissus tazetta* (flowers and bulb).

"Demulcent" is used in the treatment of boils and mastitis. The root is emetic. It is used to relieve headaches. The chopped root is applied externally as an anti-phlogistic and analgesic poultice to abscesses, boils, and other skin complaints. The plant has a folklore of effectiveness against certain forms of cancer. This might be due to benzaldehyde (C_7H_6O) changing to laetrile-like compounds or to lycorine ($C_{16}H_{17}NO_4$) changing to lycobetaine-like compounds in the body (Zargari, 2014; Mozaffarian, 2011; pfaf.org; Habib *et al.*, 2007).

Orchis maculata (O. fuchsii, O. longibracteata, Dactylorhiza maculata, D. fuchsii, D. longibracteata) (Figure 5.15) is a bulb that grows up to a size of 0.6 m. The species is hermaphrodite and pollinated by bees and beetles. It has a preference for light (sandy), medium (loamy), and heavy (clay) soils. It can grow in acidic soils and can tolerate very acid soils. It can grow in semi-shade (light woodland) or no shade. It prefers moist soil.

"Salep" is very nutritive and demulcent. It has been used as a diet of special value for children and convalescents, being boiled with water, flavored, and prepared in the same way as arrowroot. Rich in mucilage, it forms a soothing and demulcent jelly that is used in the treatment of irritations of the gastro-intestinal canal. One part of salep to 50 parts water is sufficient to make a jelly. The tuber, from which salep is prepared, should be harvested as the plant dies down after flowering and setting seed. In industry, it is used for ice cream making (Zargari, 2014; Mozaffarian, 2011; pfaf.org; Mozaffarian, 1996; Da Silva, 2013).

Oryza sativa (Figure 5.16) is an annual that grows to a size of 1.8 m by 0.3 m at a fast rate. It has a preference for light (sandy), medium (loamy), and heavy (clay) soils. It can grow in acidic, neutral, and basic (alkaline) soils. It cannot grow in the

FIGURE 5.15 *Orchis maculata* (whole plant and leaves).

FIGURE 5.16 *Oryza sativa* (whole plant, spikelets, and grains).

shade. It prefers moist or wet soil and can grow in water. "Rice" is a nutritive, sooth-ing, tonic herb that is diuretic, reduces lactation, improves digestion, and controls sweating. The seeds are taken internally in the treatment of urinary dysfunction. The seeds, or the germinated seeds, are taken to treat excessive lactation. The ger-minated seeds are used to treat poor appetite, indigestion, abdominal discomfort, and bloating. The grains are often cooked with herbs to make a medicinal gruel. The rhizome is taken internally in the treatment of night sweats, especially in cases of tuberculosis and chronic pneumonia. The rhizomes are harvested at the end of the growing season and dried for use in decoctions. In industry, it is used for making ethanol, as well as animal and bird food (Zargari, 2014; Mozaffarian, 2011; pfaf.org; Nantiyakul *et al.*, 2012).

Phoenix dactylifera (Figure 5.17) is a tree that grows up to a size of 15 m at a slow rate. The flowers are hermaphrodite. It has a preference for light (sandy), medium (loamy), and heavy (clay) soils. It can grow in acidic, neutral, and basic (alkaline) soils, as well as in semi-shade (light woodland) or no shade. It prefers moist soil.

"Date palm" contains tannins with antioxidant properties that are proven to have anti-inflammatory characteristics. Dates are also a great source of vitamin A which helps with vision, skin health, and protects from oral and lung cancer. Dates protect against age-related macular degeneration as a result of the high level of the carotenoid zeaxanthin ($C_{40}H_{56}O_2$) (Zargari, 2014; Mozaffarian, 2011; pfaf.org; Adeosun, 2016).

Phragmites communis (Ph. australis) (Figure 5.18) is a perennial that grows to a size of 3.6 m by 3 m at a fast rate. The species is hermaphrodite and pollinated by wind. It has a preference for light (sandy), medium (loamy), and heavy (clay) soils. It can grow in acidic, neutral, and basic (alkaline) soils, and can tolerate very alkaline and saline soils. It can also grow in semi-shade (light woodland) or no shade. It pre-fers moist or wet soil and can grow in water; it can also tolerate maritime exposure.

The leaves of the "common reed" are used in the treatment of bronchitis and chol-era. The ash of the leaves is applied to foul sores. A decoction of the flowers is used in the treatment of cholera and food poisoning. The ashes are styptic. The stem is anti-dote, anti-emetic, anti-pyretic, and refrigerant. The root is anti-asthmatic, anti-emetic,

Glucose: $C_6H_{12}O_6$ Zeaxanthin: $C_{40}H_{56}O_2$

FIGURE 5.17   *Phoenix dactylifera* (tree, fruits, and leaves) and its main components: Glucose, zeaxanthin.

FIGURE 5.18 *Phragmites communis* (whole plant, flowers, and leaves).

anti-pyretic, anti-tussive, depurative, diuretic, febrifuge, lithontripic, sedative, siala-gogue, and stomachic. It is taken internally in the treatment of diarrhea, fevers, vomiting, coughs with thick dark phlegm, lung abscesses, urinary tract infections, and food poisoning (especially from seafood). Externally, it is mixed with gypsum and used to treat halitosis and toothache. The root is harvested in the autumn and juiced or dried for use in decoctions. In industry, it is used for making paper and chaff (Zargari, 2014; Mozaffarian, 2011; pfaf.org; Mozaffarian, 1996; Omondi & Omondi, 2015).

Ruscus aculeatus (Figure 5.19) is an evergreen bushy tree that grows to a size of 0.8 m by 1 m at a slow rate. The species is dioecious and pollinated by insects. The plant is not self-fertile. It has a preference for light (sandy), medium (loamy), and well-drained soil, and can tolerate heavy clay and nutritionally poor soils. It can grow

Ruscogenin: $C_{27}H_{42}O_4$ Neoruscogenin: $C_{27}H_{40}O_4$

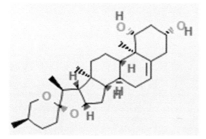
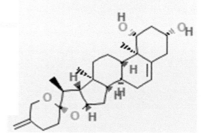

FIGURE 5.19 *Ruscus aculeatus* (shrub and fruit) and its main components: Ruscogenin, neoruscogenin.

in acidic, neutral, and basic (alkaline) soils, as well as very alkaline soils. It can also grow in full shade (deep woodland) or semi-shade (light woodland). It prefers dry or moist soil and can tolerate drought.

"Butcher's broom" is little used in modern herbalism but, in view of its positive effect upon varicose veins and hemorrhoids, it could be due for a revival. The root is aperient, deobstruent, depurative, diaphoretic, diuretic, and vasoconstrictor. It has been taken internally in the past in the treatment of jaundice, gout, and kidney and bladder stones; at the present time it is used to treat venous insufficiency and hemorrhoids. It should not be prescribed for patients with hypertension. It is applied externally in the treatment of hemorrhoids. The root is harvested in the autumn and dried for later use. The whole plant is also sometimes used. This remedy should not be given to people with high blood pressure. The plant contains saponin glycosides, including ruscogenin ($C_{27}H_{42}O_4$) and neoruscogenin ($C_{27}H_{40}O_4$). These substances are anti-inflammatory and cause the contraction of blood vessels, especially veins (Zargari, 2014; Mozaffarian, 2011; pfaf.org; Elsohly *et al.*, 1975).

Saccharum officinarum (Figure 5.20) is a perennial that grows to a size of 6 m by 1.5 m at a fast rate. It has a preference for light (sandy), medium (loamy), heavy (clay), and well-drained soil. It can grow in acidic, neutral, and basic (alkaline) soils, and can tolerate very acidic and saline soils. It cannot grow in the shade. It prefers moist or wet soil.

The leaf ash of "sugarcane" is used to treat sore eyes. The stem juice is used to treat sore throats. The sweet juice in the stem is used to treat snake bites and wounds from poison arrows. Mixed with an infusion of "wallaba" (*Eperua sp.*), it is used to treat curare poisoning. A decoction of the young leaves is used to treat urinary conditions. In industry, it is used for making animal and bird food, paper, ethanol, sugar, and enzymes (Zargari, 2014; Mozaffarian, 2011; pfaf.org; Uchenna *et al.*, 2015).

Secale cereale (Figure 5.21) is an annual that grows to a size of 1.8 m by 0.1 m. It is not frost tender. The flowers are hermaphrodite and pollinated by wind. It has a

FIGURE 5.20   *Saccharum officinarum* (whole plant, flowers, and stems).

FIGURE 5.21   *Secale cereale* (whole plant, spikelets, and grains).

preference for light (sandy), medium (loamy), heavy (clay), and well-drained soil. It can grow in acidic, neutral, and basic (alkaline) soils. It cannot grow in the shade. It prefers moist soil and can tolerate drought. The plant can tolerate strong winds but not maritime exposure.

The seed of "rye" is made into a poultice and applied to tumors. The seed is also an effective laxative due to its fibrous seed coat. It is used for the treatment of high blood pressure and diabetes. The buds are tonic (Zargari, 2014; Mozaffarian, 2011; pfaf.org; Nystrom *et al.*, 2008).

Triticum sp. (Figure 5.22) is an annual that grows up to a size of 1.3 m. It is not frost tender. The flowers are hermaphrodite and pollinated by wind. It has a preference for light (sandy), medium (loamy), heavy (clay), and well-drained soil. It can grow in acidic, neutral, and basic (alkaline) soils. It cannot grow in the shade. It prefers moist soil.

The buds of "wheat" are tonic. The flour of grains is used a lot for making bread and has an anti-poison property. Also, it is useful for the treatment of asthma and inflammation. Wheat has many uses: as a biomass for fuel, etc., for thatching, as a mulch in the garden, etc. A fiber obtained from the stems is used for making paper. The stems are harvested in late summer after the seed has been harvested; they are cut into usable pieces and soaked in clear water for 24 hours. They are then cooked for 2 hours in lye or soda ash and then beaten in a ball mill for 1.5 hours in a ball mill. The fibers make a green-tan paper. The starch from the seed is used for laundering, sizing textiles, etc. It can also be converted to alcohol for use as a fuel. In industry, it is used for making paper, chaff, as well as animal and bird food (Zargari, 2014; Mozaffarian, 2011; pfaf.org; Rajoria *et al.*, 2015; Desai *et al.*, 2015).

Zea mays (Figure 5.23) is a fast-growing annual reaching up to 2 m in size. The species is monoecious and pollinated by wind. It has a preference for light (sandy), medium (loamy), heavy (clay), and well-drained soil. It can grow in acidic and neutral soils. It cannot grow in the shade. It prefers moist soil.

A decoction of the leaves and roots of "corn" is used in the treatment of strangury, dysuria, and gravel. The corn silks are cholagogue, demulcent, diuretic, lithontripic, mildly stimulant, and vasodilator. They also act to reduce blood sugar levels and so are used in the treatment of diabetes mellitus as well as cystitis, gonorrhea, gout, etc. The silks are harvested before pollination occurs and are best used when fresh because they tend to lose their diuretic effect when stored and also become purgative. A decoction of the cob is used in the treatment of nose bleeds and menorrhagia. The seed is diuretic and a mild stimulant. It is a good emollient poultice for ulcers, swellings, and rheumatic pains, and is widely used in the treatment of cancer, tumors, and warts. It contains the cell-proliferant and wound-healing substance allantoin

FIGURE 5.22 *Triticum sp.* (whole plant, spikelets, and grains).

Cerebronic acid: $C_{24}H_{48}O_3$ Allantoin: $C_4H_6N_4O_3$

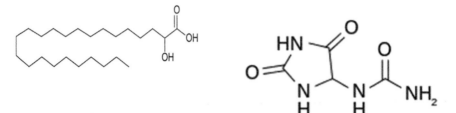

FIGURE 5.23 *Zea mays* (whole plant, fruits, and silks) and its main components: Cerebronic acid, allantoin.

($C_4H_6N_4O_3$), which is widely used in herbal medicine (especially from the herb com-frey, *Symphytum officinale*) to speed up the healing process. The plant is said to have anti-cancer properties and is experimentally hypoglycemic and hypotensive. In industry, it is used for making paper, as well as animal and bird food (Zargari, 2014; Mozaffarian, 2011; pfaf.org; Bhaigyabati *et al.*, 2011).

6

Spermatophytes: Gymnosperms

Spermatophytes are split into several sections. This chapter reviews gymnosperms which are listed in Table 6.1. A total of six species is provided and described in detail below. As each species is presented, information on the taxonomy, plant use, and specifically the key natural products isolated from each plant is provided.

In this group, all species have flowers and ovules; after fertilization, the ovules produce seeds. They have naked ovules that are not in the ovary. Most species in this group do not have a perianth. Another name of this group is pinophyta (Azadbakht, 2000).

Diversity of this group in Iran is very low, but it consists of one important family, cupressaceae, which has important uses.

Cupressus sempervirens (Figure 6.1) is an evergreen tree that grows to a size of 30 m by 5 m at a medium rate. The flowers are monoecious and pollinated by wind. It has a preference for light (sandy), medium (loamy), heavy (clay), and well-drained soil, and can tolerate nutritionally poor soil. It can grow in acidic, neutral, and basic (alkaline) soils. It cannot grow in the shade. It prefers dry or moist soil and can tolerate drought.

The cones and young branches of the "Mediterranean cypress" are anthelmintic, anti-pyretic, anti-rheumatic, antiseptic, astringent, balsamic, and vasoconstrictive. They are harvested in late winter and early spring, and then dried for later use. Taken internally, it is used in the treatment of whooping cough, the spitting up of blood, spasmodic coughs, colds, flu, and sore throats. Applied externally as a lotion or as a diluted essential oil (using an oil such as almond), it astringes varicose veins and hemorrhoids, tightening up the blood vessels. A foot bath of the cones is used to cleanse the feet and counter excessive sweating. The extracted essential oil should not be taken internally without professional guidance. A resin is obtained from the tree by making incisions in the trunk. This has a vulnerary action on slow-healing wounds and also encourages whitlows to come to a head. An essential oil from the leaves and cones is used in aromatherapy. Its keyword is "astringent" (Zargari, 2014; Mozaffarian, 2011; pfaf.org; Khabir *et al.*, 1987).

Ephedra major (Figure 6.2) is an evergreen bushy tree that grows to a size of 2 m by 1 m. The species is dioecious. The plant is not self-fertile. It has a preference for light (sandy), medium (loamy), and well-drained soil. It can grow in acidic, neutral, and basic (alkaline) soils. It cannot grow in the shade. It prefers dry or moist soil and can tolerate drought.

"Joint pine," and other members of this genus, contain various medicinally active alkaloids (but notably ephedrine ($C_{10}H_{15}NO$)) and they are widely used in preparations for the treatment of asthma and catarrh. This species is the richest source of ephedrine in India, with the stems containing over 2.5% total alkaloids, of which about 75% is ephedrine ($C_{10}H_{15}NO$). The whole plant can be used at much lower concentrations than the isolated constituents—unlike using the isolated ephedrine,

TABLE 6.1

Spermatophytes: Gymnosperms

Family Name	Species Name	Major Features of Botany	Usable Parts	Some Important Substances	Edible, Some Industrial and Medicinal Benefits
Cupressaceae	*Cupressus sempervirens*	Tree—Evergreen—Very Small and Triangular Leaf—Smooth and Red-Grey Trunk	Wood—Cone—Branch—Leaf	Tannins—Essences—Resin	Vasoconstrictor—Anthelmintic—Astringent—Balsamic—Vasoconstrictive—Treating Diarrhea, Hemorrhoids, and Bleeding—Anti-Pyretic, Anti-Rheumatic, and Antiseptic
Cupressaceae	*Juniperus communis*	Bushy Tree—Evergreen—Lots of Twigs—Rough and Dark Green Leaf	Fruit	Juniperin—Formic Acid—Acetic Acid	Tonic—Dialysis—Aromatic—Carminative—Diaphoretic—Diuretic—Rubefacient—Stomachic—Treating Cystitis, Digestive Problems, Chronic Arthritis, Gout, and Rheumatic—Anti-Inflammatory and Antiseptic
Cupressaceae	*Juniperus sabina*	Bushy Tree—Evergreen—Small Leaf	Shoot—Leaf	Sabinol—Sabinene—Cadinene—Terpenoids	Industrial (Insecticide Making)—Abortifacient—Diuretic—Emetic—Emmenagogue—Irritant—Treating Wart—Stomach and Intestine Irritation
Cupressaceae	*Thuja orientalis*	Tree—Evergreen—Very Small Leaf	Leaf—Seed—Stem—Bark of Root	Pinene—Caryophyllene—Borneol—Bornyl Acetate—Thujone—Camphor—Sesquiterpenes—Rhodoxanthin—Amentoflavone—Quercetin—Myricetin—Carotene—Xanthophyll—Vitamin C	Industrial (Dye Making)—Astringent—Diuretic—Emmenagogue—Vasoconstrictor—Emollient—Expectorant—Febrifuge—Heemostatic—Refrigerant—Stomachic—Treating Rheumatism, Cough, Diarrhea, Palpitations, Insomnia, Nervous Disorders, Constipation, and Bleeding—Anti-Bacterial, Anti-Pyretic, and Anti-Tussive

(Continued)

TABLE 6.1 (CONTINUED)

Spermatophytes: Gymnosperms

Family Name	Species Name	Major Features of Botany	Usable Parts	Some Important Substances	Edible, Some Industrial and Medicinal Benefits
Ephedraceae	*Ephedra major*	Bushy Tree—Evergreen—Thin and Small Leaf	Whole Plant Especially Stem	Ephedrine—Pseudo-Ephedrine—Alkaloids	Vasoconstrictor—Increases Blood Pressure—Diaphoretic—Diuretic—Febrifuge—Hypertensive—Nervine—Pectoral—Tonic—Vasodilator—Treating Asthma and Obesity—Anti-Viral
Taxaceae	*Taxus baccata*	Tree—Evergreen—Lots of Twigs	Leaf—Shoot—Whole Plant Except Fleshy Fruit	Taxine—Taxicatin—Taxol	Industrial (Insecticide Making)—Diuretic—Laxative—Cardiotonic—Diaphoretic—Emmenagogue—Expectorant—Narcotic—Purgative—Treating Coughs, Cystitis, Eruptions, Headaches, Rheumatism, and Asthma—Anti-Spasmodic and Anti-Cancer

FIGURE 6.1   *Cupressus sempervirens* (tree, fruits, and leaves).

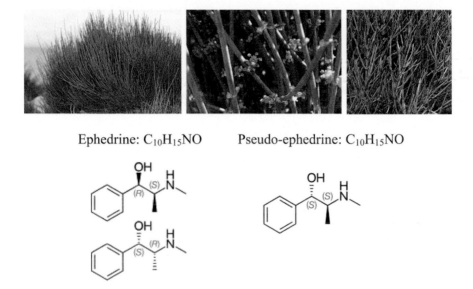

Ephedrine: $C_{10}H_{15}NO$ Pseudo-ephedrine: $C_{10}H_{15}NO$

FIGURE 6.2   *Ephedra major* (shrub, flowers, and leaves) and its main components: Ephedrine, pseudo-ephedrine.

using the whole plant rarely gives rise to side effects. The plant also has antiviral effects, particularly against influenza. The stems are a pungent, bitter, warm herb that dilates the bronchial vessels whilst stimulating the heart and central nervous system. The stems are also diaphoretic, diuretic, febrifuge, hypertensive, nervine, pectoral, tonic, vasoconstrictor, and vasodilator. They are used internally in the treatment of asthma, hay fever, and allergic complaints. They are also combined with a number of other herbs and used in treating a wide range of complaints. This plant should be used with great caution, preferably under the supervision of a quali-fied practitioner. It should not be prescribed to patients who are taking monoamine oxidase inhibitors, or suffering from high blood pressure, hyperthyroidism, or glau-coma. Ephedrine is seen as a performance-boosting herb and, as such, is a forbidden substance in many sporting events such as athletics. The stems can be harvested at any time of the year and are dried for later use (Zargari, 2014; Mozaffarian, 2011; pfaf.org; Ibragic & Sofic, 2015).

Juniperus communis (Figure 6.3) is an evergreen bushy tree that grows to a size of 9 m by 4 m at a slow rate. The species is dioecious and pollinated by wind. The plant is not self-fertile. It has a preference for light (sandy), medium (loamy), and well-drained soil, and can tolerate heavy clay and nutritionally poor soils. It can grow in acidic, neutral, and basic (alkaline) soils, as well as very acidic and very alkaline soils. It can also grow in semi-shade (light woodland) or no shade. It prefers dry or moist soil and can tolerate drought. The plant can tolerate maritime exposure.

"Juniper" fruits are commonly used in herbal medicine, as a household remedy, and also in some commercial preparations. They are especially useful in the treatment of digestive disorders as well as kidney and bladder problems. The fully ripe fruits are strongly antiseptic, aromatic, carminative, diaphoretic, strongly diuretic, rubefacient, stomachic, and tonic. They are used in the treatment of cystitis, digestive problems, chronic arthritis, gout, and rheumatic conditions. They can be eaten raw or used in a tea, but some caution is advised as large doses can irritate the urinary passage. Externally, it is applied as a diluted essential oil, having a slightly warming effect upon the skin and is thought to promote the removal of waste products from underlying tissues. It is, therefore, helpful when applied to arthritic joints, etc. The fruits should not be used internally by pregnant women as this can cause an abortion. The fruits also increase menstrual bleeding so should not be used by women with heavy periods. When made into an ointment, the fruits are applied to exposed wounds and prevent irritation by flies. The essential oil is used in aromatherapy. Its keyword is "toxin elimination" (Zargari, 2014; Mozaffarian, 2011; pfaf.org; Han *et al.*, 2017).

Juniperus sabina (Figure 6.4) is an evergreen bushy tree that grows to a size of 4 m by 4 m at a slow rate. The species is dioecious and pollinated by wind. The plant is not self-fertile. It has a preference for light (sandy), medium (loamy), heavy (clay), and well-drained soil. It can grow in acidic, neutral, and basic (alkaline) soils, and can tolerate very alkaline soils. It cannot grow in the shade. It prefers dry or moist soil and can tolerate drought. The plant can tolerate maritime exposure.

The young shoots of "savin" are abortifacient, diuretic, emetic, powerfully emmenagogue, and irritant. The plant is rarely used internally but is useful as an ointment

Acetic acid: CH₃COOH

FIGURE 6.3 *Juniperus communis* (shrub, fruits, and leaves) and its main components: Acetic acid, and formic acid (not shown).

Sabinol: $C_{10}H_{16}O$

FIGURE 6.4 *Juniperus sabina* (shrub, fruits, and leaves) and a main component: Sabinol.

and dressing to blisters, etc., in order to promote discharge. The powdered leaves are also used in the treatment of warts. The shoots are harvested in spring and dried for later use. Use with great caution and never during pregnancy (Zargari, 2014; Mozaffarian, 2011; pfaf.org; Pascual *et al.*, 1983).

Taxus baccata (Figure 6.5) is an evergreen tree that grows to a size of 15 m by 10 m at a slow rate. The species is dioecious and pollinated by wind. The plant is not self-fertile. It is noted for attracting wildlife. It has a preference for light (sandy), medium (loamy), and well-drained soil, and can tolerate heavy clay soil. It can grow in acidic, neutral, and basic (alkaline) soils, as well as in very acidic and very alkaline soils. It can also grow in full shade (deep woodland), semi-shade (light woodland), or no shade. It prefers dry or moist soil and can tolerate drought. The plant can tolerate strong winds but not maritime exposure. It can tolerate atmospheric pollution.

The "yew" tree is a highly toxic plant that has occasionally been used medicinally, mainly in the treatment of chest complaints. Modern research has shown that the plants contain the substance taxol ($C_{47}H_{51}NO_{14}$) in their shoots. Taxol has shown exciting potential as an anti-cancer drug, particularly in the treatment of ovarian cancers. Unfortunately, the concentrations of taxol ($C_{47}H_{51}NO_{14}$) in this species are too low to be of much value commercially, though it is being used for research purposes. This remedy should be used with great caution and only under the supervision of a qualified practitioner. All parts of the plant, except the fleshy fruit, are anti-spasmodic, cardiotonic, diaphoretic, emmenagogue, expectorant, narcotic, and purgative. The leaves have been used internally in the treatment of asthma, bronchitis, hiccups, indigestion, rheumatism, and epilepsy. Externally, the leaves have been used in a steam bath as a treatment for rheumatism. A homeopathic remedy is made from the young shoots and the berries, and is used in the treatment of many diseases including cystitis, eruptions, headaches, heart and kidney problems, rheumatism, etc. Ingestion of 50–100 g of needles can cause death (Zargari, 2014; Mozaffarian, 2011; pfaf.org; Thomas & Polwart, 2003).

Thuja orientalis (Figure 6.6) is an evergreen tree that grows to a size of 15 m by 5 m at a slow rate. The species is monoecious and pollinated by wind. It has a

Taxine (Taxine alkaloid): $C_{35}H_{47}NO_{10}$ Taxicatin: $C_{14}H_{20}O_8$

(Taxine A)

Taxol (Paclitaxel): $C_{47}H_{51}NO_{14}$

FIGURE 6.5 *Taxus baccata* (tree, fruits, and leaves) and its main components: Taxine, taxicatin, taxol.

preference for light (sandy), medium (loamy), heavy (clay), and well-drained soil. It can grow in acidic, neutral, and basic (alkaline) soils, and can tolerate very alkaline soils, semi-shade (light woodland), or no shade. It prefers dry or moist soil and can tolerate drought. It can tolerate atmospheric pollution.

"Platycladus" is commonly used in Chinese herbalism, where it is considered to be one of the 50 fundamental herbs. Both the leaves and the seeds contain an essential oil consisting of borneol ($C_{10}H_{18}O$), bornyl acetate ($C_{12}H_{20}O_2$), thujone ($C_{10}H_{16}O$), camphor ($C_{10}H_{16}O$), and sesquiterpenes. The leaves also contain rhodoxanthin ($C_{40}H_{50}O_2$), amentoflavone ($C_{30}H_{18}O_{10}$), quercetin ($C_{15}H_{10}O_7$), myricetin ($C_{15}H_{10}O_8$), carotene ($C_{40}H_x$), xanthophyll ($C_{40}H_{56}O_2$), and vitamin C ($C_6H_8O_6$). The leaves are antibacterial, antipyretic, antitussive, astringent, diuretic, emmenagogue, emollient, expectorant, febrifuge, hemostatic, refrigerant, and stomachic. Their use is said to improve the growth of hair. They are used internally in the treatment of coughs, hemorrhages, excessive menstruation, bronchitis, asthma, skin infections, mumps, bacterial dysentery, arthritic pain, and premature baldness. The leaves are harvested for use as required and can be used fresh or dried.

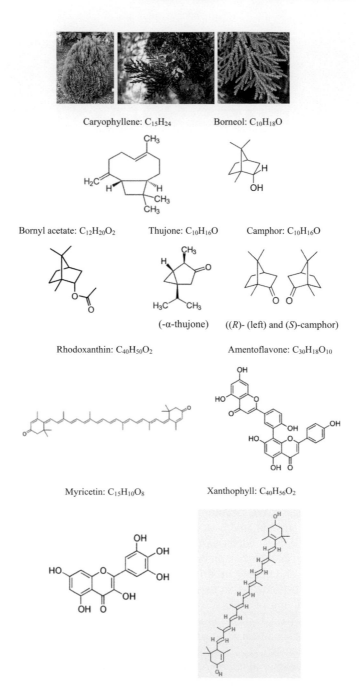

FIGURE 6.6 *Thuja orientalis* (tree, fruits, and leaves) and its main components: Caryophyllene, borneol, bornyl acetate, thujone, camphor, rhodoxanthin, amentoflavone, myricetin, xanthophyll.

This remedy should not be prescribed to pregnant women. The seed is aperient, lenitive, and sedative. It is used internally in the treatment of palpitations, insomnia, nervous disorders, and constipation in the elderly. The root bark is used in the treatment of burns and scalds. The stems are used in the treatment of coughs, colds, dysentery, rheumatism, and parasitic skin diseases (Zargari, 2014; Mozaffarian, 2011; pfaf.org; Jain & Sharma, 2017).

7

Cryptogamaes—Pteridophytes

Cryptogamaes are split into two sections. This chapter reviews pteridophytes which are listed in Table 7.1. A total of eight species is provided and described in detail below. As each species is presented, information on the taxonomy, plant use, and specifically the key natural products isolated from each plant is provided.

In this group, all species have stem, leaf, and root, but do not have flower and seed (Azadbakht, 2000).

Diversity of this group in Iran is very low, but several of these have important uses.

Adiantum capillus-veneris (Figure 7.1) is a fern that grows to a size of 0.3 m by 0.3 m at a slow rate. Preference for light (sandy), medium (loamy), and heavy (clay) soils, and well-drained soil. It can grow in neutral and basic (alkaline) soils, also semi-shade (light woodland). It prefers moist soil. "Maidenhair fern" has a long history of medicinal use and was the main ingredient of a popular cough syrup called "capillaire," which remained in use until the nineteenth century. The plant is little used in modern herbalism. The fresh or dried leafy fronds have the following properties: Anti-dandruff, anti-tussive, astringent, demulcent, depurative, emetic, weakly emmenagogue, emollient, weakly expectorant, febrifuge, galactogogue, laxative, pectoral, refrigerant, stimulant, sudorific, and tonic. A tea or syrup is used in the treatment of coughs, throat afflictions, and bronchitis. It is also used as a detoxicant in alcoholism and to expel worms from the body. Externally, it is used as a poultice on snake bites, bee stings, etc. The plant is best used fresh, though it can also be harvested in the summer and dried for later use (Zargari, 2014; Mozaffarian, 2011; pfaf.org; Dehdari & Hajimehdipoor, 2018).

Athyrium filix-femina (Figure 7.2) is a deciduous fern that grows to a size of 0.6 m by 0.5 m at a medium rate. Preference for light (sandy), medium (loamy), and heavy (clay) soils, also well-drained soil and can tolerate heavy clay soil. It can grow in acidic, neutral, and basic (alkaline) soils and can grow in very acidic soils, also full shade (deep woodland) or semi-shade (light woodland). It prefers moist soil.

A tea of the boiled stems of "lady fern" has been used to relieve labor pains. The young unfurled fronds have been eaten to treat internal ailments such as cancer of the womb. The roots are anthelmintic and diuretic. A tea of the boiled roots has been used to treat general body pains, to stop breast pains caused by childbirth, and to induce milk flow in caked breasts. The dried powdered root has been applied externally to heal sores. A liquid extract of the root is an effective anthelmintic, though it is less powerful than the male fern, *Dryopteris filix-mas* (Zargari, 2014; Mozaffarian, 2011; pfaf.org; Soare *et al.*, 2012).

Dryopteris filix-mas (Figure 7.3) is an evergreen fern that grows to a size of 1.2 m by 1 m at a medium rate. It is in leaf all year. Preference for light (sandy), medium (loamy), and heavy (clay) soils, and can tolerate nutritionally poor soil. It can

TABLE 7.1

Cryptogamaes—Pteridophytes

Family Name	Species Name	Major Features of Botany	Usable Parts	Some Important Substances	Edible, Some Industrial and Medicinal Benefits
Aspleniaceae	*Phyllitis scolopendrium* (*Asplenium scolopendrium*)	Fern—Evergreen—Short Rhizome—Leaf with Trichome	Frond	Terpenoids	Astringent—Cholagogue—Diaphoretic—Diuretic—Expectorant—Vulnerary—Treating Diarrhea, Dysentery, Cough, Kidney and Bladder Diseases
Athyriaceae	*Athyrium filix-femina*	Fern—With Rhizome—One Meter and More Height	Rhizome—Stem—Unfurled Frond	Filicic Acid—Polystichine	Anthelmintic—Diuretic—Anthelmintic—Treating Body Pains and Breast pains—Anti-Worm
Dennstaedtiaceae (hypolepidaceae)	*Preridium aquilinum*	Fern—Running Tall Horizontal Rhizome—Leaf with Trichome	Rhizome—Leaf—Shoot	Thiaminase—Coumarin—Carcinogens—Glycosides	Industrial (Adhesive, Compost Mulch, and Hair Dye Making)—Digestive—Diuretic—Refrigerant—Vermifuge—Appetizer—Tonic—Treating Cancer, Diarrhea, Colds, Arthritis, and Tuberculosis—Anti-Worm, Anti-Emetic, and Antiseptic
Dryopteridaceae (aspidiaceae)	*Dryopteris filix-mas*	Fern—Evergreen—Laminar Rhizome	Rhizome Stalk	Resin—Phytosterols—Filicic Acid—Albaspidin—Filicin	Industrial (Compost Making)—Astringent—Anodyne—Astringent—Febrifuge—Vermifuge—Vulnerary— Treating Inflammation, Hemorrhage, and Rheumatism—Anti- Inflammatory and Anti-Worm Especially Anti-Taenia, Anti-Bacterial, and Anti-Viral

(Continued)

TABLE 7.1 (CONTINUED)

Cryptogamaes—Pteridophytes

Family Name	Species Name	Major Features of Botany	Usable Parts	Some Important Substances	Edible, Some Industrial and Medicinal Benefits
Equisetaceae	*Equisetum arvense*	Evergreen—With Horizontal Rhizome	Green Stem	Equisetin—Fructose—Arabinose—Equisetic Acid	Industrial (Sandpaper, Dye, Polish, and Fungicide Making)—Diuretic—Coagulator—Anodyne—Astringent—Carminative—Diaphoretic—Galactogogue—Hemostatic—Vulnerary—Anti-Hemorrhagic and Antiseptic
Ophioglossaceae	*Ophioglossum vulgatum*	Fern—Small Rhizome—Two Kinds of Leaves (Sterile and Fertile)	Leaf—Rhizome	Sapogenins	Detergent—Styptic—Emetic—Hemostatic—Vulnerary—Antiseptic—Treating Skin Rash and Ulcers, Internal Bleeding and Bruising
Polypodiaceae	*Polypodium vulgare*	Fern—Evergreen—Fleshy Horizontal Rhizome with Thin Lamina	Rhizome—Leaf	Mucilage—Starch—Resin—Polypodine	Industrial (Insecticide Making)—Alterative—Anthelmintic—Cholagogue—Demulcent—Diuretic—Expectorant—Pectoral—Purgative—Tonic—Laxative Aperient—Mucus Producing—Anti-Bile
Pteridaceae (Adiantaceae)	*Adiantum capillus-veneris*	Fern—With Running Rhizome—Small Leaf	Leafy Frond	Capillarin—Gallic Acid—Tannins—Mucilage	Astringent—Demulcent—Depurative—Emetic—Emmenagogue—Emollient—Expectorant—Febrifuge—Galactogogue—Pectoral—Refrigerant—Stimulant—Sudorific—Tonic—Sweaty—Laxative—Mucus Producing—Anti-Dandruff and Anti-Tussive

Capillarin: $C_{13}H_{10}O_2$

FIGURE 7.1 *Adiantum capillus-veneris* (whole plant, leaves, and spores) and its main components: Capillarin, and gallic acid (not shown).

Filicic acid

(Filicic acid BBB: C_3H_7 C_3H_7)

FIGURE 7.2 *Athyrium filix-femina* (whole plant, leaves, and spores) and its main component: Filicic acid.

grow in acidic and neutral soils, also semi-shade (light woodland). It prefers dry or moist soil and can tolerate drought.

"Male fern" is one of the most popular and effective treatments for tapeworms. The root stalks are anodyne, anti-bacterial, anti-inflammatory, anti-viral, astringent, febrifuge, vermifuge, and vulnerary. The root contains an oleoresin that paralyzes tapeworms and other internal parasites and has been used as a worm expellant. The active ingredient in this oleoresin is filicin ($C_{36}H_{44}O_{12}$); roots of this species contain about 1.5–2.5% filicin ($C_{36}H_{44}O_{12}$). It is one of the most effective treatments known for tapeworms—its use should be immediately followed by a non-oily purgative such as magnesium sulphate ($MgSO_4$), *Convolvulus scammonia*, or *Helleborus niger* in order to expel the worms from the body. An oily purge, such as castor oil, increases the absorption of the fern root and can be dangerous. The root is also taken internally

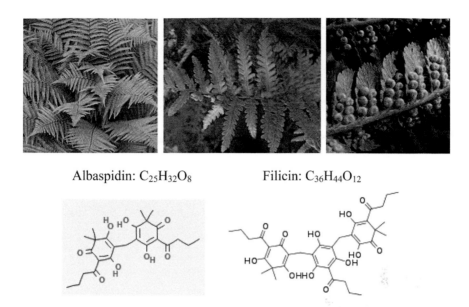

Albaspidin: $C_{25}H_{32}O_8$ Filicin: $C_{36}H_{44}O_{12}$

FIGURE 7.3 *Dryopteris filix-mas* (whole plant, leaves, and spores) and its main components: Albaspidin, filicin, and filicic acid (not shown).

in the treatment of internal hemorrhage, uterine bleeding, mumps, and feverish illnesses. The root is harvested in the autumn and can be dried for later use. This remedy should be used with caution and only under the supervision of a qualified practitioner. The root is toxic, and the dosage is critical. Pregnant women and people with heart complaints should not be prescribed this plant. See also the notes above on toxicity. Externally, the root is used as a poultice in the treatment of abscesses, boils, carbuncles, and sores. In industry, it is used for making compost (Zargari, 2014; Mozaffarian, 2011; pfaf.org; Uwumarongie, 2016).

Equisetum arvense (Figure 7.4) is an evergreen perennial that grows up to a size of 0.6 m. Preference for light (sandy), medium (loamy), and heavy (clay) soils, and can grow in nutritionally poor soil. It can grow in acidic, neutral, and basic (alkaline) soils, also semi-shade (light woodland) or no shade. It prefers dry or moist soil.

"Horsetail" has an unusual chemistry compared to most other plants. It is rich in silica (SiO_2), contain several alkaloids (including nicotine), and various minerals. Horsetail is very astringent and makes an excellent clotting agent, staunching wounds, stopping nosebleeds, and reducing the coughing up of blood. It helps speed the repair of damaged connective tissue, improving its strength and elasticity. The plant is anodyne, anti-hemorrhagic, antiseptic, astringent, carminative, diaphoretic, diuretic, galactogogue, hemostatic, and vulnerary. The green infertile stems are used; they are most active when fresh but can also be harvested in late summer and dried for later use. Sometimes the ashes of the plant are used. The plant is a useful diuretic when taken internally and is used in the treatment of kidney and bladder problems, cystitis, urethritis, prostate disease, and internal bleeding, proving especially useful when there is bleeding in the urinary tract. A decoction applied externally will stop the bleeding of wounds and promote healing. It is especially effective on nose bleeds.

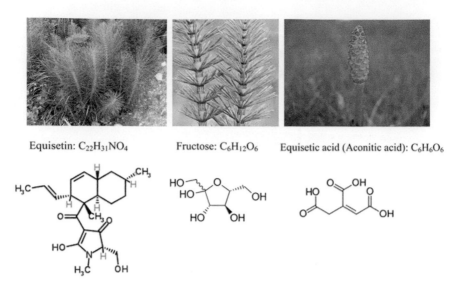

Equisetin: C$_{22}$H$_{31}$NO$_4$ Fructose: C$_6$H$_{12}$O$_6$ Equisetic acid (Aconitic acid): C$_6$H$_6$O$_6$

FIGURE 7.4 *Equisetum arvense* (whole plant, leaves, and sporangia) and its main components: Equisetin, fructose, equisetic acid, and arabinose (not shown).

A decoction of the herb added to a bath benefits slow-healing sprains and fractures, as well as certain irritable skin conditions such as eczema. The plant contains equisetic acid (C$_6$H$_6$O$_6$), which is thought to be identical to aconitic acid. This substance is a potent heart and nerve sedative that is a dangerous poison when taken in high doses. This plant contains irritant substances and should only be used for short periods of time. It is also best only used under the supervision of a qualified practitioner. A homeopathic remedy is made from the fresh plant. It is used in the treatment of cystitis and other complaints of the urinary system. In industry, it is used for making polish, dye, sandpaper, and fungicide (Zargari, 2014; Mozaffarian, 2011; pfaf.org; Al Snafi, 2017).

Ophioglossum vulgatum (Figure 7.5) is a fern that grows up to a size of 0.3 m. Preference for light (sandy), medium (loamy), and heavy (clay) soils, and well-drained soil. It can grow in acidic, neutral, and basic (alkaline) soils. It cannot grow in the shade. It prefers moist soil.

FIGURE 7.5 *Ophioglossum vulgatum* (whole plant, leaves, and spike).

The root and leaves of the "adders-tongue fern" are antiseptic, detergent, emetic, hemostatic, styptic, and vulnerary. An ointment made from the plant is a good remedy for wounds and is also used in the treatment of skin ulcers. The expressed juice of the leaves is drunk as a treatment for internal bleeding and bruising (Zargari, 2014; Mozaffarian, 2011; pfaf.org; Minarchenko *et al.*, 2017).

Phyllitis scolopendrium (Asplenium scolopendrium) (Figure 7.6) is an evergreen fern that grows to a size of 0.6 m by 0.5 m at a slow rate. It is in leaf all year. Preference for light (sandy) and medium (loamy) soils, and can tolerate heavy clay soil. It can grow in acidic, neutral, and basic (alkaline) soils, and also very alkaline soils. It can also grow in full shade (deep woodland) or semi-shade (light woodland). It prefers dry or moist soil.

The fronds of the "hart's-tongue fern" are astringent, cholagogue, diaphoretic, diuretic, expectorant, and vulnerary. Externally it is used as an ointment in the treatment of piles, burns, and scalds. An infusion is taken internally for the treatment of diarrhea, dysentery, gravelly deposits of the bladder, and for removing obstructions of the liver and spleen. The fronds are harvested during the summer and can be dried for later use (Zargari, 2014; Mozaffarian, 2011; pfaf.org; Mozaffarian, 1996; Irudayaraj & Johnson, 2011).

Polypodium vulgare (Figure 7.7) is an evergreen fern that grows to a size of 0.3 m by 0.3 m at a fast rate. It is in leaf all year. Preference for light (sandy), medium (loamy), and heavy (clay) soils, well-drained soil, and can tolerate heavy clay and nutritionally poor soils. It can grow in acidic, neutral, and basic (alkaline) soils, also full shade (deep woodland) or semi-shade (light woodland). It prefers dry or moist soil.

"Polypody" stimulates bile secretion and is a gentle laxative. It should not be used externally as it can cause skin rashes. The root has the following properties: alterative, anthelmintic, cholagogue, demulcent, diuretic, expectorant, pectoral, purgative, and tonic. The leaves can also be used but are less active. A tea made from the roots is used in the treatment of pleurisy, hives, sore throats and stomach aches, and as a mild laxative for children. It was also considered of value for lung ailments and liver diseases. The poulticed root is applied to inflammations. A tea or syrup of the whole plant is anthelmintic. In industry, it is used for making insecticide (Zargari, 2014; Mozaffarian, 2011; pfaf.org; Grzybek, 1976).

Pteridium aquilinum (Figure 7.8) is a fern that grows to a size of 1.2 m by 2 m at a fast rate. Preference for light (sandy), medium (loamy), and heavy (clay) soils. It can grow in acidic and neutral soils and can tolerate very acidic soils. It can also grow in semi-shade (light woodland) or no shade. It prefers dry or moist soil. The plant can tolerate maritime exposure.

FIGURE 7.6 *Phyllitis scolopendrium* (whole plant, leaves, and spores).

Polypodine

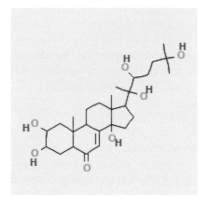

(Polypodine A: $C_{27}H_{44}O_7$)

FIGURE 7.7 *Polypodium vulgare* (whole plant, leaves, and spores) and its main components: Polypodine, and starch (not shown).

FIGURE 7.8 *Pteridium aquilinum* (whole plant, leaves, and spores).

The young shoots of the "eagle fern" are diuretic, refrigerant, and vermifuge. They have been eaten as a treatment for cancer. The leaves have been used in steam baths as a treatment for arthritis. A decoction of the plant has been used in the treatment of tuberculosis. A poultice of the pounded fronds and leaves has been used to treat sores of any type and also to bind broken bones in place. The root has the following properties: anti-emetic, antiseptic, appetizer, and tonic. A tincture of the root in wine is used in the treatment of rheumatism. A tea made from the roots is used in the treatment of stomach cramps, chest pains, internal bleeding, diarrhea, colds, and also to expel worms. The poulticed root is applied to sores, burns, and caked breasts. In industry, it is used for making hair dye, mulch, compost, and adhesive (Zargari, 2014; Mozaffarian, 2011; pfaf.org; Sunday, 2015).

8

Cryptogamaes—Non-Vascular Plants

Cryptogamaes are split into two sections. This chapter reviews non-vascular plants which are listed in Table 8.1. A total of nine species is provided and described in detail below. As each species is presented, information on the taxonomy, plant use, and specifically the key natural products isolated from each plant is provided.

In this group, all species have no stem, leaf, or root. In fact, they are plants without a vascular system. This group is equal to thallophytes recorded in other plant taxonomy methods (Azadbakht, 2000).

Diversity of this group in Iran is very low, but several of these have important uses.

Agaricus bisporus (Figure 8.1) is a fungus and its stipe holds a flat pileus. The pileus is fleshy and flat; it is also white to brown in color.

"Common mushroom" contains vitamin B_2 ($C_{17}H_{20}N_4O_6$), vitamin B_3 ($C_6H_5NO_2$), vitamin B_5 ($C_9H_{17}NO_5$), hydrazine (N_2H_4), agaritine ($C_{12}H_{17}N_3O_4$), and gyromitrin ($C_4H_8N_2O$), used for treating cancer, migraines, prostate issues, and it is also tonic (Bhushan & Kulshreshtha, 2018; Dhamodharan & Mirunalini, 2013).

Dictyota dichotoma (Figure 8.2) is a marine brown macro-alga with flat and leaf-like thallus and olive to brown in color.

"Forkweed" contains diterpenes, auxins, cytokinins, flavonoids, tannins, and coumarin ($C_9H_6O_2$). It is also used for treating cancer and inflammation (Sasikumar *et al.*, 2011; Shelar *et al.*, 2012; Deyab *et al.*, 2016).

Dunaliella salina (Figure 8.3) is a green micro-alga which is found in sea salt fields. It has two flagella of equal length.

"Green alga" is known for its antioxidant activity because of its ability to create a large amount of carotenoids. It is used in cosmetics and dietary supplements. It contains carotene ($C_{40}H_x$), zeaxanthin ($C_{40}H_{56}O_2$), and lutein ($C_{40}H_{56}O_2$), and also is anti-bacterial and in addition is useful for treating inflammation, cancer, and asthma (Helena *et al.*, 2016; Lin *et al.*, 2017).

Glypholecia scabra (Acarospora scabra) (Figure 8.4) is a lichenized fungus with a white to blue

color in dry situations and red to brown in wet. It is crustose with growth on stones.

It contains norstictic acid ($C_{18}H_{12}O_9$), fatty acids, and gyrophic acid ($C_{24}H_{20}O_{10}$). It is also used for its anti-bacterial properties (Shafiee *et al.*, 2016; Younesi *et al.*, 2015).

Gracilaria corticata (Figure 8.5) is a marine red macro-alga with thallus that consists of a bundle of flat and much divided blades.

It contains agaroid, phycocolloids, and fatty acids. It is diuretic, anti-bacterial, anti-viral, and anti-fungal, and also is useful for treating arthritis, cancer, and inflammation (Paul J, 2014; Smit, 2004; De Almeida *et al.*, 2011; Gharangik *et al.*, 2018).

TABLE 8.1

Cryptogamaes—Non-Vascular Plants

Family Name	Species Name	Major Features of Botany	Usable Parts	Some Important Substances	Edible, Some Industrial and Medicinal Benefits
Acarosporaceae	*Glypholecia scabra (Acarospora scabra)*	Lichenized Fungi—White to Blue When Dry and Red to Brown When Wet—Crustose	Whole Plant	Norstictic Acid—Gyrophoric Acid—Fatty Acids	Anti-Bacterial
Agaricaceae	*Agaricus bisporus*	Fungi—Stipe Holds the Spread Pileus—Flat and Fleshy Pileus and White to Brown in Color	Whole Plant	Vitamin B_2, B_3, and B_5 Hydrazine—Agaritine—Gyromitrin	Edible—Tonic—Treating Cancer, Prostate, and Migraine
Dictyotaceae	*Dictyota dichotoma*	Marine Brown Macro-Algae—Thallus Flat and Leaf-Like—Olive to Brown in Color	Whole Plant	Diterpenes—Auxins—Cytokinins—Flavonoids—Tannins—Coumarin	Treating Inflammation and Cancer
Dunaliellaceae	*Dunaliella salina*	Green Micro-Algae—Found in Sea Salt Fields—With Two Flagells of Equal Length	Whole Plant	Carotenoids—Carotene—Vitamin A—Zeaxanthin—Lutein	Industrial (Cosmetics and Dietary Supplements)—Anti-Bacterial and Anti-Oxidant—Treating Asthma, Inflammation, and Cancer
Gracilariaceae	*Gracilaria corticata*	Marine Red Macro-Alga—Thallus Consists of Bundle of Flat and Much Divided Blades	Whole Plant	Agaroid—Phycocolloids—Fatty Acids	Diuretic—Anti-Bacterial, Anti-Viral, and Anti-Fungal—Treating Arthritis, Inflammation, and Cancer
Hymenochoetaceae	*Phellinus conchatus*	Fungi—Firm and Woody or Cork-like Trama and Brown in Color	Sporocarp	Phellinsin—Hispidin	Anti-Bacterial, Anti-Viral, and Anti-Fungal—Treating Sore Throat, Cancer, and Inflammation

(Continued)

TABLE 8.1 (CONTINUED)

Cryptogamaes—Non-Vascular Plants

Family Name	Species Name	Major Features of Botany	Usable Parts	Some Important Substances	Edible, Some Industrial and Medicinal Benefits
Parmeliaceae	*Usnea hirta*	Fruticose Lichen—Elastic Chord—Growing on Tree Host—Green to Grey in Color	Whole Plant	Usnic Acid—Phenolic Acids—Fatty Acids—Flavonoids—Sterols	Industrial (Dye in Textile Making)—Antibiotic, Anti-Bacterial, Anti-Viral, and Anti-Fungal—Treating Flu, Bronchitis, Pneumonia, Sore Throats, and Skin Infections
Sargassaceae	*Sargassum ilicifolium*	Marine Brown Macro-Alga—Few Meters in Length—Dark Green to Brown in Color	Whole Plant	Fucoxanthin—Fatty Acids	Anti-Bacterial, Anti-Viral, and Anti-Fungal—Treating Inflammation and Cancer
Tuberaceae	*Tuber aestivum*	Fungi—Peridium is Brown to Black in Color and with Pyramidal Warts Similar to Hard Crust	Whole Plant	Vitamin B_1, B_2, and B_3	Edible—Stomach Tonic—Laxative—Coagulater—Treating Diarrhea and Bloody Diarrhea

Hydrazine: N_2H_4 Agaritine: $C_{12}H_{17}N_3O_4$ Gyromitrin: $C_4H_8N_2O$

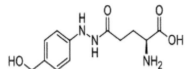

Vitamin B_5 (Pantothenic acid): $C_9H_{17}NO_5$

FIGURE 8.1 *Agaricus bisporus* (fungus and mycelium) and its main components: Hydrazine, agaritine, gyromitrin, vitamin B_5.

FIGURE 8.2 *Dictyota dichotoma* (alga and thallus).

Phellinus conchatus (Figure 8.6) is a fungus with a hard and woody or cork-like trama and is brown in color.

It contains phellinsin ($C_{18}H_{14}O_8$) and hispidin ($C_{13}H_{10}O_5$). The part which is used is the sporocarp which useful for treating cancer, sore throats, and inflammation. In addition, it is anti-bacterial, anti-viral, and anti-fungal (Hokmollahi *et al.*, 2011; Ren *et al.*, 2006).

Sargassum ilicifolium (Figure 8.7) is a marine brown macro-alga reaching up to a few meters in length and it is dark green to brown in color.

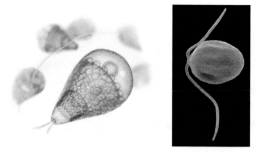

FIGURE 8.3 *Dunaliella salina* (alga and flagella).

Norstictic acid: $C_{18}H_{12}O_9$ Gyrophoric acid: $C_{24}H_{20}O_{10}$

FIGURE 8.4 *Glypholecia scabra* (lichenized fungus) and its main components: Norstictic acid, gyrophoric acid..

FIGURE 8.5 *Gracilaria corticata* (alga and thallus).

Phellinsin: $C_{18}H_{14}O_8$ Hispidin: $C_{13}H_{10}O_5$

FIGURE 8.6   *Phellinus conchatus* (fungus) and its main components: Phellinsin, hispidin.

Fucoxanthin: $C_{42}H_{58}O_6$

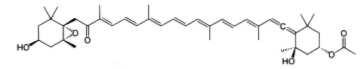

FIGURE 8.7   *Sargassum ilicifolium* (alga and thallus) and its main component: Fucoxanthin.

"Sargassum big leaves" contains fucoxanthin ($C_{42}H_{58}O_6$) and fatty acids. It is used for treating inflammation and cancer. In addition, it has anti-bacterial, anti-viral, and anti-fungal properties (Simpi *et al.*, 2013; Rebecca, 2012; Hafezieh, 2018; Etemadian *et al.*, 2017).

Tuber aestivum (Figure 8.8) is a fungus with a brown to black peridium which has pyramidal warts similar to a hard crust.

FIGURE 8.8   *Tuber aestivum* (fungus).

"Summer truffle" contains vitamin B_1 ($C_{12}H_{17}N_4OS+$), vitamin B_2 ($C_{17}H_{20}N_4O_6$), and vitamin B_3 ($C_6H_5NO_2$). It is edible and is useful for treating diarrhea and bloody diarrhea. It is a stomach tonic, laxative, and coagulater (Paolocci *et al.*, 2004; Gryndler *et al.*, 2011; Ammarellou & Alvarado, 2017).

Usnea hirta (Figure 8.9) is a fruticose lichen with a green to grey color which grows on a tree host.

"Shaggy beard lichen" contains elastic chords and usnic acid ($C_{18}H_{16}O_7$) (which is one of the most common and abundant lichen metabolites, well-known as an antibiotic), phenolic acids, fatty acids, and sterols. It is used for treating skin infections, sore throats, pneumonia, flu, and bronchitis. In addition, it is anti-bacterial, anti-viral, and anti-fungal. In industry, it is used as a dye in textile making (Singh *et al.*, 2016; Solberg, 1987).

Usnic acid: $C_{18}H_{16}O_7$

FIGURE 8.9   *Usnea hirta* (fruticose lichen) and its main component: Usnic acid.

9

Conclusion

Those who understand the power of medicinal plants believe that using processed materials rather than using the herbs directly may not achieve the plants' full therapeutic effect. Therefore, there are limitations to the use of individual chemical drugs from plants. Furthermore, if the packaging of plant material is required, there may be limitations:

1. The farther away the habitat of a medicinal plant, the harder it is to access.
2. The inability to grow any medicinal plant in any area.
3. Most of the useful plant members, even if kept under special conditions, will eventually lose their medicinal effects.
4. Cultivating plants, even with all the necessary growing conditions, may not produce a similar plant possessing all of the effective ingredients of the corresponding wild species.

In 2019, exports of medicinal plants from Iran reached $520 million. Of course, as mentioned in the introduction, given the special geographic location of Iran and its different climatic conditions, there is a large variety of medicinal plants, so more production and exports can be expected. The global trade of medicinal plants exceeds $107 billion in 2019. Iran ranks first in the world in the export of *Crocus sativus*, *Rosa damascena*, *Frula gummosa*, *Astragalus sp.*, and *Carum carvi* products. 190 species of wild medicinal plants exist in the country within the rangelands which have comparative therapeutic advantages. In addition, many kinds of medicinal plants are produced in greenhouses in Iran under supervision of the Ministry of Agriculture.

On average, about 220 species of medicinal plants are used in traditional Iranian medicine. More than 17.2 million kilograms of medicinal plants are produced annually by 209 cooperatives in Iran. Of course, some of the rangelands used by local people should also be added to this figure; however, no exact statistics are available (irna.ir).

The Fars province produces the most medicinal plants in Iran, and, according to the provincial Department of Natural Resources, there are more than 1600 species of medicinal plants in the province and more than 100 species of those are largely cultivated on 600 hectares of land within the province. It should be noted that the Fars province is 133,000 km² in size, which amounts to 8.1% of the total country. It is located in the southern part of Iran and features three climates, resulting in a high diversity of medicinal plants (farsnews.com).

Regarding medicinal plants in Iran, the following can be noted:

1. Climatic variation in Iran allows for a wide range of plants from different families to grow.

2. The most abundant families of medicinal plants in Iran include leguminoseae, rosaceae, cruciferae, and umbelliferae.

3. The diversity of the dialypetalaes plants in Iran is greater than the others.

4. Many medicinal plants in Iran such as *Crocus sativus*, *Oryza sativa*, *Cerasus arium*, *Citrus limonum*, and *Pistacia vera* are found only in certain parts and their distribution is limited.

5. Some medicinal plants such as *Peganum harmala*, *Punica granatum*, *Triticum sp.*, and *Astragalus sp.* are widely distributed and have been observed in other places too.

6. The development of the production and utilization of medicinal plants is one of the most important ways of creating jobs for local people and it is only possible with the efficient cooperation of government.

7. There is yet further potential in Iran to expand production of medicinal plants and evaluate further their pharmacological effects and therapeutic benefits.

References

Abdullah, K., A.T. Mahmood, H.H. Siddiqui & J. Akhtar, 2016, Phytochemical and pharmacological properties on *Citrus limetta* (Mosambi), *Journal of Chemical and Pharmaceutical Research*, Vol. 83, pp 555–563.

Abu- Reidah, I. M., R.M. Jamous & M.S. Ali- Shtayeh, 2014, Phytochemistry, pharmacological properties and industrial application of *Rhus coriaria* L. (Sumac), *JJBS*, Vol. 7, pp 233–244.

Adeosun, A.M., S.O. Oni, O.M. Ighodaro, O.H. Durosinlorun & O.M. Oyedele, 2016, Phytochemical, minerals and free radical scavenging profiles of *Phoenix dactilyfera* L. seed extract, *Journal of Taibah University Medical Sciences*, Vol. 11(1), pp 1–6.

Adl, E. 1970, *Divisions of Climate and Vegetation of Iran*, Tehran University Press.

Ahmad Baba, Sh., A. Malik, Z. Ahmed Wani & T. Mohiuddin, 2015, Phytochemical analysis and antioxidant activity of different tissue types of *Crocus sativus* and oxidative stress alleviating potential of saffron extract in plants, bacteria, and yeast, *South African Journal of Botany*, Vol. 99, pp 80–87.

Akinyele, B.J & O.S. Oloruntoba, 2017, Comparative studies on *Citrullus vulgaris*, *Citrullus colocynthis* and *Cucumeropsis mannii* for ogiri production, *British Microbiology Research Journal*, Vol. 3(1), pp 1–18.

Al Haadi, A.H., S.S. Rahbi, S. Akhtar, S. Said, A. Weli & Q Al Riyami, 2013, Phytochemical screening, antibacterial and cytotoxic activities of *Petroselinum Crispum* leaves grown in Oman, *Iranian Journal of Pharmaceutical Sciences*, Vol. 9(1), pp 61–65.

Al Haidari, R., G. Mohamed, E. Elkhayat & M. Moustafa, 2010, Cucumol A: A cytotoxic triterpenoid from *Cucumis melo* seeds, *Revista Brasileira de Farmacognosia*, Vol. 26(6), pp 701–704.

Al Snafi, A.E., 2013, The pharmaceutical importance of *Althaea officinalis* and *Althaea rosea*, *International Journal of Pharm Tech Research*, Vol. 5(3), pp 1378–1385.

Al Snaf, A.E., 2014, Chemical constituents and pharmacological activities of *Arachis hypogaea*, *IJPRS*, Vol. 3(1), pp 615–623.

Al Snaf, A.E., 2015, The pharmacological importance of *Asparagus officinalis*, *Journal of Pharmaceutical Biology*, Vol. 5(2), pp 93–98.

Al Snafi, A.E., 2016, A review on chemical constituents and pharmacological activities of *Coriandrum sativum*, *Journal of Pharmacy*, Vol. 6(7), pp 17–42.

Al Snaf, A.E., 2016, The medical importance of *Cicer arietinum*, *Journal of Pharmacy*, Vol. 6(3), pp 29–40.

Al Snafi, A.E., 2017, The pharmacology of *Equisetum arvense*, *Journal of Pharmacy*, Vol. 7(2), pp 31–42.

Ali, M., Sh. Sultana & M. Jameel, 2016, Phytochemical investigation of flowers of *Rosa damascena* Mill, *International Journal of Herbal Medicine*, Vol. 4(6), pp 179–183.

Aliverdinia, A., & W.A. Pridemore. 2008. An overview of the illicit narcotics problem in the Islamic Republic of Iran. *European Journal of Crime. Criminal Law and Criminal Justice*, Vol. 16, pp 155–170.

Altinyay, C., I. Suntar, L. Altun, H. Keles & E. Kupeli Akkol, 2016, Phytochemical and biological studies on *Alnus glutinosa* subsp. glutinosa, *A. orientalis* var. orientalis and *A. orientalis* var. pubescens leaves, *Journal of Ethnopharmacology*, Vol. 4(192), pp 148–160.

Amiri Tehranizadeh, Z., A. Baratian & H. Hosseinzadeh, 2016, Russian olive (Elaeagnus angustifolia) as a herbal healer, *Bioimpacts*, Vol. 6(3), pp 157–167.

Ammarellou, A & P. Alvarado, 2017, First report of *Tuber aestivum* var. uncinatum from Iran based on morphological and molecular characteristics, *Rostaniha*, Vol. 18(2), pp 227–228.

Angelini, L.G, L. Pistelli, P. Belloni, A. Bertoli & S. Panconesi, 1997, *Rubia tinctorum* a source of natural dyes: Agronomic evaluation, quantitative analysis of alizarin and industrial assays, *Industrial Crops and Products*, Vol. 6(3-4), pp 303–311.

Apraj, V.D & N.S. Pandita, 2014, Pharmacognostic and phytochemical evaluation of *Citrus reticulata* blanco peel, *International Journal of Pharmacognosy and Phytochemical Research*, Vol. 6(2), pp 328–331.

Arora, E.K., V. Sharma, A. Khurana & A. Manchanda, 2017, Phytochemical analysis and evaluation of antioxidant potential of ethanol extract of *Allium cepa* and ultra-high homoeopathic dilutions available in the market: A comparative study, *Indian Journal of Research in Homoeopathy*, Vol. 11(2), pp 88–96.

Ashraf, M.U., G. Muhammad, M.A. Hussain & S.N.A. Bukhari, 2016, *Cydonia oblonga* M., a medicinal plant rich in phytonutrients for pharmaceuticals, *Front. Pharmacol.* Vol. 7(163), pp 1–20.

Aslam, M.Sh., B.A. Choudhary, M. Uzair & A.S. Ijaz, 2012, The genus Ranunculus: A phytochemical and ethnopharmacological review, *International Journal of Pharmacy and Pharmaceutical Sciences*, Vol. 4(5), pp 15–22.

Ayeni, E.A., A. Abubakar, G. Ibrahim, V. Atinga & Z. Muhammad, 2018, Phytochemical, nutraceutical and antioxidant studies of the aerial parts of *Daucus carota* L. (Apiaceae), *Journal of Herbmed Pharmacology*, Vol. 7(2), pp 68–73.

Ayoob, I., Y.M. Hazari, Sh.U. Rehman, M.M. Fazili & Kh.A. Bhat, 2017, Phytochemical and cytotoxic evaluation of *Peganum Harmala*: Structure activity relationship studies of harmine, *Chemistry Select*, Vol. 2(10), pp 2965–2968.

Ayoub, N.A., 2003, Unique phenolic carboxylic acids from *Sanguisorba minor*, *Phytochemistry*, Vol. 63, pp 433–436.

Azadbakht, M. 2000, *Medicinal Plants Taxonomy*, Tabib Press, p. 404.

Azizi, Sh., A. Ghasemi Pirbalouti & M. Amirmohammadi, 2014, Effect of Hydro-alcoholic extract of persian oak (*Quercus brantii*) in experimentally gastric ulcer, *Iranian Journal of Pharmaceutical Research*, Vol. 13(3), pp 967–974.

Badoni, R., D.K. Semwal, U. Rawat & M.S.M. Mahat, 2011, Chemical constituents from fruits and stem bark of *Celtis australis* L., *Helvetica Chimica Acta*, Vol. 94(3), pp 464–473.

Benaida, N.D., D.A. Kilani, V.B. Schini Keirth, N. Djebbli & D. Atmani, 2013, Pharmacological potential of *Populus nigra* extract as antioxidant, anti-inflammatory, cardiovascular and hepatoprotective agent, *Asian Pacific Journal of Tropical Biomedicine*, Vol. 3(9), pp 697–704.

Bergqvist, M.H.J & N.U. Olsson, 1992, Characterization of honeysuckle (*Lonicéra caprifolium* L.) seed oil triacylglycerols by high performance liquid chromatography and light scattering detection, *Phytochemical Analaysis*, Vol. 3(5), pp 215–217.

Bhaigyabati, T., T. Kirithika, J. Ramya & K. Usha, 2011, Phytochemical constituents and antioxidant activity of various extracts of corn silk (*Zea mays.* L), *Research Journal of Pharmaceutical, Biological and Chemical Sciences*, Vol. 2(4), pp 986–993.

Bhat, A.H., A. Alia, B. Kumar & S. Mubashir, 2018, Phytochemical constituents of genus nepeta, *Journal of Chemistry*, Vol. 7(2), pp 31–37.

Bhushan, A & M. Kulshreshtha, 2018, The medicinal mushroom *Agaricus bisporus*: Review of phytopharmacology and potential role in the treatment of various diseases, *Journal of Nature and Science of Medicine*, Vol. 1(1), pp 4–9.

Bina, F & R. Rahimi, 2017, Sweet marjoram: A review of ethnopharmacology, phytochemistry, and biological activities, *Journal of Evidence-Based Integrative Medicine*, Vol. 22(1), pp 175–185.

Blumenthal, M., W.R. Busse, J. Klein, American Botanical Council, R. Rister, T. Hall, C. Riggins, J. Gruenwald & A. Goldberg, 1998, *The Complete German Commission E Monographs: Therapeutic Guide to Herbal Medicines*, American Botanical Council Press, p. 685.

Bora, K.S & A. Sharma, 2011, Phytochemical and pharmacological potential of *Medicago sativa*, *Pharmaceutical Biology*, Vol. 49(2), pp 211–220.

Boroja, T., J. Katanic, G. Rosic, … & V. Mihailovic, 2018, Summer savory (*Satureja hortensis* L.) extract: Phytochemical profile and modulation of cisplatin-induced liver, renal and testicular toxicity, *Food and Chemical Toxicology*, Vol. 118, pp 252–263.

Boudaoud Ouahmed, H., S.A.T. Tiab, N. Saidani & M. Gherrou, 2015, Phytochemical screening and pharmacological activities of *Ulmus campestris* bark extracts, *Oriental Pharmacy and Experimental Medicine*, Vol. 15(4), pp 1–13.

Boulaaba, M., S. Mejd, S. Mariem & Kh. Mkadmini, 2015, Antimicrobial activities and phytochemical analysis of *Tamarix gallica* extracts, *Industrial Crops and Products*, Vol. 76(76), pp 1114–1122.

Bradic, J., A. Petkovic & M. Tomovic, 2017, Phytochemical and pharmacological properties of some species of the genus galium L. (*Galium verum* and mollugo), *SJECR*. doi:10.1515/sjecr-2017-0057.

Bylka, W., M. Szaufer Hajdrych, I. Matlawska & O. Goslinska, 2004, Antimicrobial activity of isocytisoside and extracts of *Aquilegia vulgaris* L., *Letters in Applied Microbiology*, Vol. 39(1), pp 93–97.

Carvalho, M.S.S., M.D.G. Cardoso, L.R.M. Albuquerque, … & L.F.L. Silva, 2015, Phytochemical screening, extraction of essential oils and antioxidant activity of five species of unconventional vegetables, *American Journal of Plant Sciences*, Vol. 6, pp 2632–2639.

Catarino, M.D., A.M.S. Silva, M.T. Cruz & S.M. Cardoso, 2017, Antioxidant and anti-inflammatory activities of *Geranium robertianum* L. decoctions, *Food & Function*, Vol. 8(9), pp 1–34.

Chandra Joshi, B., M. Mukhija & A.N. Kalia, 2014, Pharmacognostical review of *Urtica dioica* L., *International Journal of Green Pharmacy*, Vol. 8(4), pp 201–209.

Chantreau, M., B. Chabbert, S. Billiard, S. Hawkins & G. Neutelings, 2015, Functional analyses of cellulose synthase genes in flax (*Linum usitatissimum*) by virus-induced gene silencing, *Plant Biotechnol Journal*, Vol. 13(9), pp 1312–1324.

Chhatre, S., T. Nesari, G. Somani, D. Kanchan & S. Sathaye, 2014, Phytopharmacological overview of *Tribulus terrestris*, *Pharmacognosy Reviews*, Vol. 8(15), pp 45–51.

Chikkulla, R., S.R. Mondi & K.M. Gottumukkula, 2018, A review of *Gossypium herbaceum* (Linn), *IJPSR*, Vol. 9(9), pp 116–120.

Choudhary, D. & A. Alam, 2017, Pharmacology and phytochemistry of isoflavonoids from iris species, *Journal of Pharmacology & Clinical Research*, Vol. 3(2), pp 1–6.

Claudia, D.P., C.H. Mario, N.O. Arturo, … & P.C. Jose, 2018, Phenolic compounds in organic and aqueous extracts from *Acacia farnesiana* pods analyzed by ULPS-ESI-Q-oa/TOF-MS. In vitro antioxidant activity and anti-inflammatory response in CD-1 mice, *Molecules*, Vol. 23(9).

Colak, F., F. Savaroglu & S. Ilhan, 2009, Antibacterial and antifungal activities of *Arum maculatum*, *Journal of Applied Biological Sciences*, Vol. 3(3), pp 13–16.

Cui, X., Zh. Ma, L. Bai & Y. Wu, 2017, Phytochemical analysis of *Ziziphus jujuba* leaves in six cultivars at the whole life stage by high performance liquid chromatography, *Chemical Research in Chinese Universities*, Vol. 33(5), pp 702–708.

Czerwinska, M.E, A. Swierczewska & S. Granica, 2018, Bioactive constituents of *Lamium album* L. as inhibitors of cytokine secretion in human neutrophils, *Molecules*, Vol. 23(11), pp 1–15.

Da Silva, J.A.T., 2013, Orchids: Advances in tissue culture, genetics, phytochemistry and transgenic biotechnology, *Floriculture and Ornamental Biotechnology*, Vol. 7(1), pp 1–52.

Dall Acqua, S., R. Cervellati, E. Speroni & S. Costa, 2009, Phytochemical composition and antioxidant activity of *Laurus nobilis* L. leaf infusion, *Journal of Medicinal Food*, Vol. 12(4), pp 869–876.

Das, G & M.M. Joseph, 2017, Phytochemical screening of oats (*AVENA SATIVA*), *World Journal of Pharmaceutical Research*, Vol. 6(4), pp 1150–1155.

Dayeni, M & R. Omidbaigi, 2013, Essential oil content and constituents of *Cercis siliquastrum* L. growing in Iran, *Journal of essential oil-bearing plants JEOP*, Vol. 9(2), pp 140–143.

De Almeida, C.L.F., H.D.S. Falcao, G.R.D.M. Lima, ... & L.M. Batista, 2011, Bioactivities from marine algae of the genus gracilaria, *International Journal of Molecular Sciences*, Vol. 12(7), pp 4550–4573.

Dehdari, S & H. Hajimehdipoor, 2018, Medicinal properties of *Adiantum capillus-veneris* Linn. in traditional medicine and modern phytotherapy, *Iranian Journal of Public Health*, Vol. 47(2), pp 188–197.

Desai, S., A.H. Doddamani, R. Patil, N.B. Navalagatti & U.U. Haveri, 2015, Phytochemical and antimicrobial analysis of *Triticum aestivum* and bauhinia variegate corresponding, *Journal of Agroecology and Natural Resource Management*, Vol. 2(5), pp 394–398.

Deyab, M., T. Elkatony & F. Ward, 2016, Qualitative and quantitative analysis of phytochemical studies on brown seaweed, *Dictyota dichotoma*, *IJDER*, Vol. 4(2), pp 674–678.

Dhaliwali, H.K., R. Singh, J.K. Sidhu & J.K. Grewal, 2016, Phytopharmacological properties of *Cuminum Cyminum* linn. as a potential medicinal seeds, *World Journal of Pharmacy and Pharmaceutical Sciences*, Vol. 5(6), pp 478–489.

Dhamodharan, G & S. Mirunalini, 2013, A detail study of phytochemical screening, antioxidant potential and acute toxicity of *Agaricus bisporus* extract and its chitosan loaded nanoparticles, *Journal of Pharmacy Research*, Vol. 6(8), pp 818–822.

Dong, X., J. Fu, X. Yin, ... & J. Ni, 2014, Pharmacological and other bioactivities of the genus polygonum, *Tropical Journal of Pharmaceutical Research*, Vol. 13(10), pp 1749–1759.

Duh, P.D., G.Ch. Yen, W.J. Yen & L.W. Chang, 2001, Antioxidant effects of water extracts from barley (*Hordeum vulgare* L.) prepared under different roasting temperatures, *J. Agric. Food Chem.*, Vol. 49(3), pp 1455–1463.

El Sohly, M.A., M.M. Radwan, W. Gul, S. Chandra & A. Galal, 2017, Phytochemistry of *Cannabis sativa* L., *Phytocannabinoids*, Vol. 103, pp 1–36.

Elias, R., A.M. Diaz Lanza, E. Vidal Ollivier, G. Balansard, R. Faure & A. Babadjamian, 1991, Triterpenoid saponins from the leaves of *Hedera helix*, *Journal of Natural Products*, 54 (1), pp 98–103.

Elsohly, M.A., J.E. Knapp, K.F. Slatkin & P. Schiff, 1975, Constituents of *Ruscus aculeatus*, *Lioydia*, Vol. 38(2), pp 106–108.

Ercisli, S & E. Orhan, 2007, L. (Moraceae): Phytochemistry, traditional uses and biological activities, *Food Chemistry*, Vol. 103(4), pp 1380–1384.

Ercisli, S., M. Sengul, H. Yildiz & S. Sener, 2012, Phytochemical and antioxidant characteristics of medlar fruits (*Mespilus germanica* L.), *Journal of Applied Botany and Food Quality*, Vol. 85(1), pp 86–90.

Esalat Nejad, H & A. Esalat Nejad, 2013, *Rubia tinctorum* L. (Rubiaceae) or Madder as one of the living color to dyeing wool, *Iternational Journal of Advanced Biological and Biomedical Research*, Vol. 1(11), pp 1315–1319.

Etemadian, Y., A.A. Khanipour, B. Shabanpour, ... & Y. Zahmatkesh, 2017, Investigating the pharmacological properties of 5 native brown algae in the southern waters of the country and promoting them for breeding, *Advanced Aquaculture Sciences Journal*, Vol. 1(1), pp 43–52.

Fars News Agency. http://www.farsnews.com.

Fattorusso, E., V. Lanzotti, O. Taglialatela Scafati & C. Cicala, 2001, The flavonoids of leek, *Allium porrum*, *Phytochemistry*, Vol. 57(4), pp 565–569.

Garachh, D., M.A. Patel, M. Chakraborty & J.V. Kamath, 2012, Phytochemical and pharmacological profile of *Punica granatum*, *International Research Journal of Pharmacy*, Vol. 3(2), pp 65–68.

Gharangik, B.M., E. Kamrani & M. Kokabi, 2018, Comparison of two species of native agarophytes on the coast of southern Iran for aquaculture: Gracilaria corticata (J.Ag.) J.Ag. &*Gracilariopsis longissima* (S.G. Gmelin) M. Steentoft, L. M.Irvine & W.F. Farnham, *Journal of Aquatic Animals Ecology*, Vol. 6(4), pp 136–140.

Gryndler, M., H. Hrselova, L. Soukupova & E. Streiblova, 2011, Detection of summer truffle (*Tuber aestivum* Vittad.) in ectomycorrhizae and in soil using specific primers, *FEMS Microbiology Letters*, Vol. 318(1), pp 84–91.

Grzybek, J., 1976, Biological and phytochemical investigations on *Polypodium vulgare* growing in Poland part 3 poly phenolic compounds and free sugars in leaves, *Acta Biologica Cracoviensia Series Botanica*, Vol. 19(2), pp 69–82.

Habib, A.H., N. Abdel Azim, Kh. Shams & N.M. Hassan, 2007, Evaluation of *Narcissus tazetta* under different habitats, *Asian Journal of Chemistry*, Vol. 19(6), pp 1–7.

Hafezieh, M., 2018, Antioxidant properties of phenol compounds in brown seaweed, *Sargassum ilicifolium* of Iranian shore coast of Oman sea, *ISFJ*, Vol. 26(5), pp 13–22.

Hakimimeybodi, M.H. 2009, *Identification of Iranian Rangeland Plants*, University Press Center, p. 189.

Hamedi, A., M.M. Zarshenas, M. Sohrabpour & A. Zargaran, 2013, Herbal medicinal oils in traditional Persian medicine, *Pharmaceutical Biology*, Vol. 51(9), pp 1208–1218.

Han, X., T.L. Parker & C. Benavente, 2017, Anti-inflammatory activity of Juniper (*Juniperus communis*) berry essential oil in human dermal fibroblasts, *Cogent Medicine*, Vol. 4(1), pp 1–7.

Haouat, A.Ch., S.E. Guendouzi, A. Haggoud, ... & M. Iraqui, 2013, Antimycobacterial activity of *Populus alba* leaf extracts, *Journal of Medicinal Plants Research*, Vol. 7(16), pp 1015–1021.

Hasplova, K., A. Hudecova, E. Miadokova & Z. Magdolenova, 2011, Biological activity of plant extract isolated from *Papaver rhoeas* on human lymfoblastoid cell line, *Neoplasma*, Vol. 58(5), pp 386–391.

Hayat, M.Gh., 2013, Phytochemical analysis of *Nigella sativa* and its antibacterial activity against clinical isolates identified by ribotyping, *International Journal of Agriculture and Biology*, Vol. 15(6), pp 1151–1156.

Helena, Sh., M. Zainuri & J. Suprijanto, 2016, Microalgae *Dunaliella salina* (Teodoresco, 1905) growth using the LED light (light limiting dioda) and different media, *Aquatic Procedia*, Vol. 7, pp 226–230.

Hennia, A., M.G. Miguel & S. Nemmiche, 2018, Antioxidant activity of *Myrtus communis* L. and *Myrtus nivellei* batt. & trab. extracts, *Medicines (Basel)*, Vol. 5(3), pp 89–96.

Henning, W & K. Herrmann, 1980, Flavonol tetraglycosides from the leaves of *Prunus cerasus* and *P. Avium*, *Phytochemistry*, Vol. 19(12), pp 2727–2729.

Hokmollahi, F., H. Rafati, H. Riahi, ... & S. Mosazade, 2011, Collection and identification of a medicinal mushroom, *Phellinus Conchatus* in Iran and investigation of the antibacterial activity of total methanol extract and fractional extracts, *Journal of Shaheed Sadoughi University of Medical Sciences*, Vol. 18(6), pp 521–530.

Hosseinpour Jaghdani, F., T. Shomali, S. Gholipour Shahraki, M. Rahimi Madiseh & M. Rafiean Kopaei, 2017, *Cornus mas*: A review on traditional uses and pharmacological properties, *Journal of Complementary and Integrative Medicine*, Vol. 14(3).

Hosseinzadeh, H., S.A. Sajadi Tabassi, N. Milani Moghadam, M. Rashedinia & S. Mehri, 2012, Antioxidant activity of *Pistacia vera* fruits, leaves and gum extract, *IJPR*, Vol. 11(3), pp 879–887.

Huang, N., L. Rizshsky, C. Hauck, B.J. Nikolau, P.A. Murphy & D.F. Birt, 2011, Identification of anti-inflammatory constituents in *Hypericum perforatum* and *Hypericum gentianoides* extracts using RAW 264.7 mouse macrophages, *Phytochemistry*, Vol. 72(16), pp 2015–2023.

Hussain, T., I. Ahmad Baba, S.M. Jain & A. Bashir, 2013, Phytochemical screening of methanolic extract of *Prunus Persica*, *International Journal of Scientific Research*, Vol. 4(3), pp 52–53.

Ibragic, S & E. Sofic, 2015, Chemical composition of various Ephedra species, *Bosnian Journal of Basic Medical Sciences*, Vol. 15(3), pp 21–27.

Imam, M.Z & S. Akter, 2011, *Musa paradisiaca* L. and *Musa sapientum* L.: A phytochemical and pharmacological review, *Journal of Applied Pharmaceutical Science*, Vol. 1(5), pp 14–20.

Irudayaraj, V & M. Johnson, 2011, Pharmacognostical Studies On Three Aspleniumspecies, *Journal of Phytology*, Vol. 3(10), pp 1–9.

Islamic Republic News Agency. http://www.irna.ir.

Jafari Foutami, I., M. Akbarlou, A. Sepehry, M. Mazandarani & M.R. Forouzeh, 2018, An investigation of phytochemical, total phenolic and flavonoid contents and antioxidant activity in aerial parts of two species of salvia and the effect of environmental factors on their distribution in Behshahr Hezarjarib area, *Herbal Medicine*, Vol. 4(1), pp 1–9.

Jafari, M & A. Tavili, 2012, *Reclamation of Arid Lands*, Tehran University Press, p. 396.

Jain, N & M. Sharma, 2017, Ethanobotany, phytochemical and pharmacological aspects of *Thuja orientalis*, *Indian Journal of Pure & Applied Biosciences*, Vol. 5(3), pp 73–83.

Jameel, M., A. Ali & M. Ali, 2014, New phytoconstituents from the aerial parts of *Fumaria parviflora* Lam, *Journal of Advanced Pharmaceutical Technology & Research*, Vol. 5(2), pp 64–69.

Jaradat, N.A., A.N. Zaid & F. Hussein, 2016, Investigation of the anti-obesity and antioxidant properties of wild *Plumbago europaea* and *Plumbago auriculata* from North Palestine, *Chemical and Biological Technologies in Agriculture*, Vol. 3(31), pp 1–9.

Jena, J & A. Gupta, 2012, *Ricinus communis* linn: A phytopharmacological review, *International Journal of Pharmacy and Pharmaceutical Sciences*, Vol. 4(4), pp 25–29.

Jones, W.T & J.L. Mangan, 1977, Complexes of the condensed tannins of sainfoin (*Onobrychis viciifolia* scop.) with fraction 1 leaf protein and with submaxillary mucoprotein, and their reversal by polyethylene glycol and pH, *Journal of the Science of Food and Agriculture*, Vol. 28(2), pp 126–136.

Jouri, M.H & M. Mahdavi, 2010, *Applied Identification of Rangeland Plants*, Aeezh Press, p. 436.

Juan, W., W. Bi Hua, C. Li, W. Wei, H. Xi Gui, X. Zheng Jun & Z. You Liang, 2012, Comparative analysis of essential oils from the leaves of wild and domestic Chimonanthus praecox, *Journal of Medicinal Plants Research*, Vol. 6(14), pp 2832–2838.

Kampuss, K & H.L. Pedersen, 2003, A review of red and white currant (*Ribes rubrum* L.), *Small Fruits Review*, Vol. 2(3), pp 23–46.

Kapoor, I.P., B. Singh, G. Singh, ... & C.A. Catalan, 2010, Chemistry and antioxidant activity of essential oil and oleoresins of black caraway (*Carum bulbocastanum*) fruits, *Journal of the Science of Food and Agriculture*, Vol. 90(3), pp 385–390.

Kaur, G.J & D.S. Arora, 2009, Antibacterial and phytochemical screening of *Anethum graveolens*, *Foeniculum vulgare* and *Trachyspermum ammi*, *BMC Complementary and Alternative Medicine*, Vol. 9(30), pp 1–10.

Kaur, R. & V. Arya, 2012, Ethnomedicinal and phytochemical perspectives of *Pyrus communis* Linn., *Journal of Pharmacognosy and Phytochemistry*, Vol. 1(2), pp 14–19.

Kaur, T., K. Hussain, S. Koul, R. Vishwakarma & D. Vyas, 2013, Evaluation of nutritional and antioxidant states of *Lepidium latifolium* Linn.: A novel phytofood from Ladakh, *PLoS One*, Vol. 8(8), e69112.

Khabir, M., F. Khatoon & W.H. Ansari, 1987, Flavonoids of *Cupressus sempervirens* and *Cupressus cashmeriana*, *Journal of Natural Products*, Vol. 50(3), pp 511–512.

Khair-ul-Bariyah, S., D. Ahmed & M. Ikram, 2012, *Ocimum Basilicum*: A review on phytochemical and pharmacological studies, *Pakistan Journal of Chemistry*, Vol. 2(2), pp 78–85.

Khan, Abdullah A., T. Mahmood, H.H. Siddiqui & J. Akhtar, 2016, Phytochemical and pharmacological properties on *Citrus limetta* (Mosambi), *Journal of Chemical and Pharmaceutical Research*, Vol. 83, pp 555–563.

Kim, K.H., E. Moon, S.R. Lee, K.J. Park, S.Y. Kim, S.U. Choi & K.R. Lee, 2015, Chemical constituents of the seeds of *Raphanus sativus* and their biological activity, *Journal of the Brazilian Chemical Society*, vol. 26(11), pp 2307–2312.

Kocheki Shahmokhtar, M & S. Armand, 2017, Phytochemical and biological studies of fennel (*Foeniculum vulgare Mill.*) from the South West Region of Iran (Yasouj), *Natural Product Chemistry and Research*, Vol. 5(4), pp 1–4.

Kooti, W & N. Daraei, 2017, A review of the antioxidant activity of celery (*Apium graveolens* L), *Journal of Evidence-Based Complementary & Alternative Medicine*, Vol. 22(4), pp 1029–1034.

Kosheleva, L.I & G.K. Nikonov, 1968, Phytochemical study of *Daphne mezereum* L., *Farmatsiia*, Vol. 17(6), pp 40–47.

Kremer, D., I. Kosalec, M. Locatelli, ... & M.Z. Koncic, 2011, Anthraquinone profiles, antioxidant and antimicrobial properties of *Frangula rupestris* (Scop.) Schur and *Frangula alnus* Mill. Bark, *Food Chemistry*, Vol. 131(4), pp 1174–1180.

Krishnaveni, M & S. Saranya, 2016, Phytochemical characterization of *Brrasica nigra* seeds, *International Journal of Advanced Life Sciences*, pp 150–159.

Kumar, R & R. Patwa, 2018, Study of phytochemical constituents and anti-oxidant activity of *Spinacia oleracea* L. of Bundelkhand Region, *IJLSSR*, pp 1599–1604.

Laghari, A.H., A.A. Memon, Sh. Memon & A. Nelofar, 2010, Determination of free phenolic acids and antioxidant capacity of methanolic extracts obtained from leaves and flowers of camel thorn (*Alhagi maurorum*), *Natural Product Research*, Vol. 26(2), pp 173–176.

Liberal, J., V. Francisco, G. Costa & A. Figueirinha, 2014, Bioactivity of *Fragaria vesca* leaves through inflammation, proteasome and autophagy modulation, *Journal of Ethnopharmacology*, Vol. 158(2), pp 113–122.

Lin, H.W., Ch.W. Liu, D.J. Yang, … & J. Ka, 2017, *Dunaliella salina* alga extract inhibits the production of interleukin-6, nitric oxide, and reactive oxygen species by regulating nuclear factor-κB/Janus kinase/signal transducer and activator of transcription in virus-infected RAW264.7 cells, *Journal of Food and Drug Analysis*, Vol. 25(4), pp 908–918.

Ma, Y.T., S.C. Hsiao, H.F. Chen & F.L. Chen, 1997, Tannins from *Albizia lebbek*, *Phytochemistry*, Vol. 46(8), pp 1451–1452.

Maggioni, L., R. Von Bothmer, G. Poulsen & E. Lipman, 2018, Domestication, diversity and use of *Brassica oleracea* L., based on ancient Greek and Latin texts, *Genetic Resources and Crop Evolution*, Vol. 65(1), pp 137–159.

Mahtout, R., V.M. Ortiz Martinez, M.J. Salar Garcia, … & L.J. Lozano Blanco, 2018, Algerian carob tree products: A comprehensive valorization analysis and future prospects, *Sustainability*, Vol. 10(1), pp 90–100.

Majewski, M., 2014, *Allium sativum*: Facts and myths regarding human health, Roczniki Państwowego Zakładu Higieny, Vol. 65(1), pp 1–8.

Majidi, Z & S.N. Sadati Lamardi, 2018, Phytochemistry and biological activities of *Heracleum persicum*, Journal of Integrative Medicine, Vol. 16(4), pp 223–235.

Maps of the world- Large detailed political map of Iran with roads, cities and airports. http://www.vidiani.com.

Mathew, B.B., S.K. Jatawa & A. Tiwari, 2012, Phytochemical analysis of citrus limonum pulp and peel, *International Journal of Pharmacy and Pharmaceutical Sciences*, Vol. 4(2), pp 369–371.

Mawa, Sh., Kh. Husain & I. Jantan, 2013, *Ficus carica L. (Moraceae): Phytochemistry, Traditional Uses and Biological Activities*, Hindawi Publishing Corporation, Article ID 974256.

Mazimba, O., R.R.T. Majinda & D. Motlhanka, 2011, Antioxidant and antibacterial constituents from *Morus nigra*, *African Journal of Pharmacy and Pharmacology*, Vol. 5(6), pp 751–754.

Mesdaghi, M. 2010, *Range Management in Iran*, Imam Reza University Press, p. 336.

Minarchenko, V., I. Tymchenko, T. Dvirna & L. Makhynia, 2017, A review of the medicinal ferns of Ukraine, *Scripta Scientifa Pharmaceutica*, Vol. 4(1), pp 7–23.

Mittal, P., V. Gupta, M. Goswami, N. Thakur & P. Bansal, 2015, Phytochemical and pharmacological potential of *Viola odorata*, *International Journal of Pharmacognosy*, Vol. 2(5), pp 215–220.

Mohiyuddin Khan, R, W. Ahmad, M. Ahmad & A. Hasan, 2016, Phytochemical and pharmacological properties of *Carum carvi*, *European Journal of Pharmaceutical and Medical Research*, Vol. 3(6), pp 231–236.

Monika, H., H. Flachowsky, M. Viola Hanke & V. Semenov, 2013, Assessment of phenotypic variation of *Malus orientalis* in the North Caucasus region, *Genetic Resources and Crop Evolution*, Vol. 60(4), pp 1463–1477.

Moosavi, Kh.S., S. Hosseini, Gh. Dehghan & A. Jahanban Esfahlan, 2014, The effect of gamma irradiation on phytochemical content and antioxidant activity of stored and none stored almond (*Amygdalus communis* L.) hull, Pharmaceutical Sciences, Vol. 20(3), pp 102–106.

Moradi Kor, N., M.B. Didarshetaban & H.R. Saeid Pour, 2013, Fenugreek (*Trigonella foenum-graecum* L.) as valuable medicinal plant, *International journal of Advanced Biological and Biomedical Research*, Vol. 1(8), pp 922–931.

Moradpour, M., S. Hafez Ghoran & J. Asghari, 2017, Phytochemical investigation of *Melissa officinalis* L. flowers from Northern part of Iran (Kelardasht), *Journal of Medicinal Plants Studies*, Vol. 5(3), pp 176–181.

Mozaffarian, V., 1996, *A Dictionary of Iranian Plant Names*, Farhange Moaser Press, p. 756.

Mozaffarian, V. 2011, *Identification of Medicinal and Aromatic Plants of Iran*, Farhange Moaser Press, p. 1444.

Munafo Jr, J.P & T.J. Gianfagna, 2015, Chemistry and biological activity of steroidal glycosides from the Lilium genus, *Natural Product Reports*, Vol. 32(3), pp 454–477.

Nagal, M., M. Kubo, K. Takahashi, M. Fujita & T. Inoue, 1983, Studies on the constituents of aceraceae plants. v. two diarylheptanoid glycosides and an arylbutanol apiosylglucoside from acer nikoense, *Chemical and Pharmaceutical Bulletin*, Vol. 31(6), pp 1917–1922.

Najafian, Y., Sh.S. Hamedi, M. Kaboli Farshchi & Z. Fevzabadi, 2018, *Plantago major* in traditional persian medicine and modern phytotherapy: A narrative review, *Electron Physician*, Vol. 10(2), pp 6390–6399.

Nandhini, S., K.B. Narayanan & K. Ilango, 2018, *Valeriana officinalis*: A review of its traditional uses, phytochemistry and pharmacology, *Asian Journal of Pharmaceutical and Clinical Research*, Vol. 11(1), pp 36–41.

Nantiyakul, N., S. Furse, I. Fisk & T.J. Foster, 2012, Phytochemical composition of *Oryza sativa* (Rice) bran oil bodies in crude and purified isolates, *Journal of American Oil Chemists Society*, Vol. 89(10), pp 1867–1872.

Naval, N.D & T.N. Pandya, 2011, Pharmacognostic study of Lepidium sativum Linn. (Chandrashura), *Ayu*, Vol. 32(1), pp 116–119.

Nazaruk, J & P. Orlikowski, 2016, Phytochemical profile and therapeutic potential of *Viscum album* L., *Natural Product Research*, Vol. 30(4), pp 373–385.

News network of medicinal plants of Iran. http://www.medplant.ir.

Niknam, F., A. Azadi, A. Barzegar, P. Faridi, N. Tanideh & M.M. Zarshenas, 2016, Phytochemistry and Phytotherapeutic Aspects of *Elaeagnus angustifolia* L., *Current Drug Discovery Technologies*, Vol. 13(4), pp 199–210.

Niknam, V & S.Y. Salehi Lisar, 2004, Chemical composition of Astragalus: Carbohydrates and mucilage content, *Pakistan Journal of Botany*, Vol. 36(2), pp 381–388.

Nimrouzi, M. & M.M. Zarshenas, 2016, Phytochemical and pharmacological aspects of *Descurainia sophia* Webb ex Pranti: Modern and traditional applications, *Avicenna Journal of Phytomedicine*, Vol. 6(3), pp 266–272.

Nystrom, L., A.M. Lampi, A.A. Andersson & A. Kamal Eldin, 2008, Phytochemicals and dietary fiber components in rye varieties in the HEALTHGRAIN diversity screen, *Journal of Agricultural and Food Chemistry*, Vol. 56(21), pp 9758–9766.

Omar Bakr, R., M. El Naa, S.S. Zaghloul & M.M. Omar, 2017, Profile of bioactive compounds in *Nymphaea alba* L. leaves growing in Egypt: Hepatoprotective, antioxidant and anti-inflammatory activity, *BMC Complementary and Alternative Medicine*, Vol. 17(1), pp 1–13.

Omondi, S & J.C. Omondi, 2015, Phytochemical analysis of 50 selected plants found in the University Botanic Garden, Maseno, Kenya for their chemotaxonomic values, *Journal of Medicinal Herbs and Ethnomedicine*, Vol. 1, pp 130–135.

Orhan, I.E & M. Karta, 2011, Insights into research on phytochemistry and biological activities of *Prunus armeniaca* L. (apricot), *Food Research International*, Vol. 44(5), pp 1238–1243.

Owfi, R.E & H. Barani, 2017, Investigation of backgrounds of product, supply and use of Astragalus genus Case study: Fars province, Iran, *IJPSAT*, Vol. 11(2), pp 86–93.

Panth, N., K.R. Paudel & R. Karki, 2016, Phytochemical profile and biological activity of *Juglans regia*, *Journal of Integrative Medicine*, Vol. 14(5), pp 359–373.

Paolocci, F., A. Rubini, C. Riccioni, F. Topini & S. Arcioni, 2004, *Tuber aestivum* and *Tuber uncinatum*: Two morphotypes or two species? *FEMS Microbiology Letters*, Vol. 235(1), pp 109–115.

Pareek, A.K., Sh. Garg, M. Kumar & S.M. Yadav, 2015, *Prosopis cineraria*: A gift of nature for pharmacy, *IJPSR*, Vol. 6(6), pp 958–964.

Pascual, J., A. Feliciano, J.M.M. Del Corral & A.F. Barrero, 1983, Terpenoids from *Juniperus Sabina*, *Phytochemistry*, Vol. 22(1), pp 300–301.

Patil, S.R., R.S. Patil & A. Godghate, 2016, *Mentha Piperita* Linn: Phytochemical, antibacterial and dipterian adulticidal approach, *International Journal of Pharmacy and Pharmaceutical Sciences*, Vol. 8(3), pp 352–355.

Patra, J.K., E.S. Kim, K.H. Oh, … & K.H. Baek, 2015, Bactericidal effect of extracts and metabolites of *Robinia pseudoacacia* L. on Streptococcus mutans and porphyromonas gingivalis causing dental plaque and periodontal inflammatory diseases, *Molecules*, Vol. 20, pp 6128–6139.

Paul Das, M & S.K. Sivagnanam, 2013, Preliminary phytochemical analysis of *Illicium verum* and *Wedelia chinensis*, *International Journal of Pharm Tech Research*, Vol. 5(2), pp 324–330.

Paul, J., 2014, Screening of diuretic activity of methanol extract of *Gracilaria corticata* J.AG. (red seaweed) in hare island, Thoothukudi, Tamil Nadu, India, *American Journal of Biological and Pharmaceutical Research*, Vol. 1(2), pp 83–87.

Pawar, A.V., S.J. Patil, & S.G. Killedar, 2017, Uses of *Cassia fistula* Linn as a medicinal plant, *International Journal of Advance Research and Development*, Vol. 2(3), pp 85–91.

Petrakis, P.V., K. Spanos, A. Feest & E. Daskalakou, 2011, Phenols in leaves and bark of *Fagus sylvatica* as determinants of insect occurrences, *International Journal of Molecular Sciences*,Vol. 12, pp 2769–2782.

Piccolella, S., A. Fiorentino, S. Pacifico, B. Ð. Dficono, P. Uzzo & P. Monac, 2008, Antioxidant properties of sour cherries (*Prunus cerasus* L.): Role of colorless phytochemicals from the methanolic extract of ripe fruits, *Journal of Agricultural and Food Chemistry*, Vol. 56(6), pp 1928–1935.

Plant for a future. http://www.pfaf.org.

Politeo, O., M. Bektasevic, I. Carey, M. Jurin & M. Roje, 2018, Phytochemical composition, antioxidant potential and cholinesterase inhibition potential of extracts from *Mentha pulegium* L., chem. *Biodivers*, Vol. 15(12), pp 41–49.

Poonam, V., R. Varshney, G. Kumar & Ch.Sh. Reddy Lokasani, 2011, Chemical constituents of the genus prunus and their medicinal properties, *Current Medicinal Chemistry*, Vol.18(25), pp 3758–3824.

Popova, I. E & M.J. Morra, 2014, Simultaneous quantification of sinigrin, sinalbin, and anionic glucosinolate hydrolysis products in *Brassica juncea* and *Sinapis alba* seed extracts using ion chromatography, *Journal of Agricultural and Food Chemistry*, Vol. 62(44), pp 10687–10693.

Raghavendra, M.P., S. Satish & K.A. Raveesha, 2006, Phytochemical analysis and antibacterial activity of *Oxalis corniculata*; a known medicinal plant, *My Science*, Vol. 1(1), pp 72–78.

Rahimi Madiseh, M., Z. Lorigoini, H. Zamani Gharaghoshi & M. Rafiean Kopaei, 2017, *Berberis vulgaris*: Specification and traditional uses, *The Iranian Journal of Basic Medical Sciences*, 20(5), pp 569–587.

Rajoria, A., A. Mehta, P. Mehta, L. Ahirwal & S. Shukla, 2015, Phytochemical analysis and estimation of major bioactive compounds from *Triticum aestivum* L. grass with antimicrobial potential, *Pakistan Journal of Pharmaceutical Sciences*, Vol. 28(6), pp 2221–2225.

Raudone, L., R. Raudonis, K. Gaivelyte & A. Pukalskas, 2014, Phytochemical and antioxidant profiles of leaves from different Sorbus L. species, *Natural Product Research*, Vol. 29(3), pp 1–5.

Rebecca, J., 2012, Antibacterial activity of *Sargassum ilicifolium* and *Kappaphycus alvarezii*, *Journal of Chemical and Pharmaceutical Research*, Vol. 4(1), pp 700–705.

Ren, G., X.Y. Liu, H.K. Zhu, S.Z. Yang & C.X. Fu, 2006, Evaluation of cytotoxic activities of some medicinal polypore fungi from China, *Fitoterapia*, Vol. 77(5), pp 408–410.

Rhee, M.H., H.J. Park & J.Y. Cho, 2009, *Salicornia herbacea*: Botanical, chemical and pharmacological review of halophyte marsh plant, *Journal of Medicinal Plants Research*, Vol. 3(8), pp 548–555.

Rojo, M.A., L. Citores, P. Jimenez, … & T. Girbes, 2003, Isolation and characterization of a new d-galactose-binding lectin from *Sambucus racemosa* L., *Protein & Peptide Letters*, Vol. 10(3), pp 287–293.

Saadat Talab, T., S. Pirveisiani & M.R. Delnavazi, 2017, Phytochemical constituents, antioxidant activity and toxicity potential of the essential oil from *Ferula gummosa* Boiss. Roots, *Research Journal of Pharmacognosy*, Vol. 4, pp 23–28.

Saberi Zafarghandi, M.B., M. Jadidi, N. Khalili. 2015. Iran's activities on prevention, treatment and harm reduction of drug abuse, *International Journal of High Risk Behaviors and Addiction*, 4(4), e22863.

Saeidnia, S & A.R. Gohari, 2012, Importance of *Brassica napus* as a medicinal food plant, *Journal of Medicinal Plant Research*, Vol. 6(14), pp 2700–2703.

Safder, M., N. Riaz, M. Imran & H. Nawaz, 2009, Phytochemical Studies on *Asphodelus tenuifolius*, *Journal- Chemical Society of Pakistan*, Vol. 31(1), pp 122–125.

Salama, H.M.H & N. Marraiki, 2009, Antimicrobial activity and phytochemical analyses of *Polygonum aviculare* L. (Polygonaceae), naturally growing in Egypt, *Saudi Journal of Biological Sciences*, Vol. 17(1), pp 57–63.

Saleem, U., K. Hussain, M. Ahmad, N.I. Bukhari, A. Malik & B. Ahmad, 2014, Report: Physicochemical and phytochemical analysis of *Euphorbia helioscopia* (L.), *Pakistan Journal of Pharmaceutical Sciences*, Vol. 27(3), pp 577–585.

Sarangowa, O., T. Kanazawa, M. Nishizawa, … & T. Yamagishi, 2014, Flavonol glycosides in the petal of *Rosa* species as chemotaxonomic markers, *Phytochemistry*, Vol. 107, pp 61–68.

Sarmamy, A.O.I., 2018, Effects of aqueous and ethanol extracts of Akaka plants *Allium Akaka* Gmel on some standard pathogenic bacteria, *Kufa Journal for Agricultural Sciences*, Vol. 10(3), pp 100–120.

Sasikumar, K., T. Govindanand & C. Anuradha, 2011, Effect of seaweed liquid fertilizer of *Dictyota dichotoma* on growth and yield of *Abelmoschus esculantus* L., *European Journal of Experimental Biology*, Vol. 1(3), pp 223–227.

Schadler, V & S. Dergatschewa, 2017, *Rubus caesius* L. leaves: Pharmacognostic analysis and the study of hypoglycemic activity, *Journal of Physiology, Pharmacy and Pharmacology*, Vol. 7(5), pp 501–508.

Schlereth, A., C. Becker, C. Horstmann, J. Tiedemann & K. Muntz, 2000, Comparison of globulin mobilization and cysteine proteinases in embryonic axes and cotyledons during germination and seedling growth of vetch (*Vicia sativa* L.), *Journal of Experimental Botany*, Vol. 51(349), pp 1423–1433.

Sebastian, P., H. Schaefer, I.R.H. Telford & S.S. Renner, 2010, Cucumber (*Cucumis sativus*) and melon (*C. melo*) have numerous wild relatives in Asia and Australia, and the sister species of melon is from Australia, *PNAS*, Vol. 107(32), pp 14269–14273.

Senthamil Selvi, R., R.Z.A. Kumar & A. Bhaskar, 2016, Phytochemical investigation and in vitro antioxidant activity of *Citrus sinensis* peel extract, *Der Pharmacia Lettre*, Vol. 8(3), pp 159–165.

Sever Yilmaz, B., G. Saltan Citoglu, M.L. Altun & H. Ozbek, 2007, Antinociceptive and Anti-inflammatory Activities of *Viburnum lantana*, *Pharmaceutical Biology*, Vol. 45(3), pp 653–658.

Shafi, S & N. Tabassum, 2015, Preliminary phytochemical screening, renal and haematological effects of *Portulaca oleracea* (Whole Plant) in swiss albino mice, *International Research Journal of Pharmacy*, Vol. 6(6), pp 349–353.

Shafiee, M., J. Hemmat & M.R. Sam, 2016, Evaluating the antibacterial activity of acetone extract of *Glypholecia scabra* and Rhizoplaca melanophthalm on some standard bacteria, *Micro Organisms Biology (Isfahan University)*, Vol. 19(5), pp 127–136.

Shahidi, F., C. Alasalyar & Ch.M. Liyana Pathirana, 2007, Antioxidant phytochemicals in hazelnut kernel (*Corylus avellana* L.) and hazelnut byproducts, *Journal of Agricultural and Food Chemistry*, Vol. 55(4), pp 1212–1220.

Shakeri, A., N. Hazeri, J. Valizadeh, A. Ghasemi & F. Zaker Tavallaie, 2014, Phytochemical screening, antimicrobial and antioxidant activities of *Anabasis aphylla* extracts, *Kragujevac Journal of Science*, Vol. 34(34), pp 71–78.

Sheikh, N.A., T. Desai & R.D. Patel, 2016, Pharmacognostic evaluation of *Melilotus officinalis* Linn., *Pharmacognosy Journal*, Vol. 8(3), pp 239–242.

Shelar, P.S., V.K. Reddy, S.G.S. Shelar, … & V.S. Reddy, 2012, Medicinal value of seaweeds and its applications, *Continental Journal of Pharmacology and Toxicology Research*, Vol. 5(2), pp 1–22.

Shojaii, A & M. Abdollahi Fard, 2012, Review of pharmacological properties and chemical constituents of *Pimpinella anisum*, *ISRN Pharmaceutics*. doi:10.5402/2012/510795.

Shokrzadeh, M & S.S. Saeedi Saravi, 2010, The chemistry, pharmacology and clinical properties of *Sambucus ebulus*, *Journal of Medicinal Plants Research*, Vol. 4(2), pp 95–103.

Simpi, Ch.C., Ch.V. Nagathan, S.R. Karaigi & N.V. Kalvane, 2013, Evaluation of marine brown algae *Sargassum ilicifolium* extract for analgesic and anti-inflammatory activity, *Pharmacognosy Research*, Vol. 5(3), pp 146–149.

Singh, A.K., R. Bharati, N.C. Manibhushan & A. Pedapati, 2013, An assessment of faba bean (*Vicia faba* L.) current status and future prospect, *African Journal of Agricultural Research*, Vol. 8(50), pp 6634–6641.

Singh, B.N., G. Prateeksha, P. Balwant Singh & R. Baipai, 2016, The genus Usnea: A potent phytomedicine with multifarious ethnobotany, phytochemistry and pharmacology, *RSC Advances*, Vol. 6(26), pp 21672–21696.

Smit, A.J., 2004, Medicinal and pharmaceutical uses of seaweed natural products, *Journal of Applied Phycology*, Vol. 16, pp 245–262.

Soare, L.C., M. Ferdes, S. Stefanov, … & A. Paunescu, 2012, Antioxidant activity, polyphenols content and antimicrobial activity of several native pteridophytes of romania, Notulae Botanicae Horti Agrobotanici Cluj-Napoca, Vol. 40(1), pp 53–57.

Solberg, Y., 1987, Chemical constituents of the lichens *cetraria delisei, lobaria pulmo-naria, stereocaulon tomentosum* and *usnea hirta, Journal of the Hattori Botanical Laboratory,* Vol. 63, pp 357–366.

Stevens, J.F., M. Ivancic, V.L. Hsu & M.L. Deinzer, 1997, Prenylflavonoids from *Humulus lupulus, Phytochemistry,* Vol. 44(8), pp 1575–1585.

Strzelecka, H., 1965, Phytochemical studies of *Consolida regalis* S. F. Gray. I. Flowers, Acta Poloniae Pharmaceutica*Acta,* Vol. 22(5), pp 453–458.

Sunday, A., 2015, Antibacterial, phytochemical and proximate analysis of *Pteridium aqui-linum, International Journal of Research on Pharmacy and Biosciences,* Vol. 2(9), pp 1–7.

Suntar, I., I. Haroon Khan, S. Patel, R. Celano & L. Rastrelli, 2018, An overview on *Citrus aurantium* L.: Its functions as food ingredient and therapeutic agent, *Oxidative Medicine and Cellular Longevity,* Vol. 9, pp 1–12.

Thai, O.D., J. Tchoumtchoua, M. Makropoulou, … & L.A. Skaltsounis, 2016, Phytochemical study and biological evaluation of chemical constituents of *Platanus orientalis* and Platanus × acerifolia buds, *Phytochemistry,* Vol. 130, pp 170–181.

Thakur, A.K & P. Raj, 2017, Pharmacological perspective of *Glycyrrhiza glabra* Linn., *Analytical & Pharmaceutical Research,* Vol. 5(5), pp 1–5.

Thantsin, Kh., 2011, Natural colorant from *Gardenia jasminoides* Ellis (Cape jasmine), *Universities Research Journal,* Vol. 4(1), pp 65–74.

Thi Ho, G.T, 2017, Bioactive compounds in flowers and fruits of *Sambucus nigra* L., Ph.D. Thesis of Oslo University.

Thomas, P.A & A. Polwart, 2003, *Taxus baccata* L., *Journal of Ecology,* Vol. 91(3), pp 489–524.

Uchenna, E.F., O.A. Adaezel & A.Ch. Steve, 2015, Phytochemical and antimicrobial prop-erties of the aqueous ethanolic extract of *Saccharum officinarum* (Sugarcane) Bark, *Journal of Agricultural Science,* Vol. 7(10), pp 291–297.

Uckoo, R., G.K. Jayaprakasha, V.M. Balasubramaniam & B.S. Patil, 2012, Grapefruit (*Citrus paradisi* Macfad) phytochemicals composition is modulated by household processing techniques, *Journal of Food Science,* Vol. 77(9), pp 921–926.

Uma, C & K.G. Sekar, 2014, Phytochemical analysis of a folklore medicinal plant *Citrullus colocynthis* L., (bitter apple), *Journal of Pharmacognosy and Phytochemistry,* Vol. 2(6), pp 195–202.

Uribe, M., M. Dibildox, S. Malpica, … & G.Garcia Ramos, 1985, Beneficial effect of vegetable protein diet supplemented with *Psyllium* plantago in patients with hepatic encephalopathy and diabetes mellitus, *Gastroenterology,* Vol. 88, pp 901–907.

Usenik, V., F. Stampar & D. Kastelec, 2013, Phytochemicals in fruits of two *Prunus domestica* L. plum cultivars during ripening, *Journal of the Science of Food and Agriculture,* Vol. 93(3), pp 681–692.

Uwumarongie, O.H., 2016, pharmacognostic evaluation and gastrointestinal activity of *Dryopteris filix-mas* (L.) Schott (Dryopteridaceae), *Ewemen Journal of Herbal Chemistry & Pharmacology Research,* Vol. 2(1), pp 19–25.

Vasas, A., O. Orban Gyapai & J. Hohmann, 2015, The Genus Rumex: Review of tra-ditional uses, phytochemistry and pharmacology, *Journal of Ethnopharmacology,* Vol. 175, pp 198–228.

Vecchio, M.G., C. Loganes & C. Minto, 2016, Beneficial and healthy properties of eucalyptus plants: A great potential use, *The Open Agriculture Journal,* Vol. 10, pp 52–57.

Vieccelli, J.C., D.L. De Siqueira, W.M. Da Silva Bispo & L.M. Carvalho Lemos, 2016, Characterization of leaves and fruits of Mango, Revista Brasileira de Fruticultura, Vol. 38(3), pp e-193.

Vinod, K.N., K.N. Ninge Gowda & R. Sudhakar, 2010, Kinetik and adsorption studies of Indian siris (*Albizia lebbeck*) natural dye on silk, *Indian Journal of Fibre & Textile Research*, Vol. 35, pp 159–163.

Vinod, M., M.K. Singh, M. Pradhan, … & Sh. Kr, 2012, Phytochemical constituents and pharmacological activities of *Betula alba* Linn., *International Journal of Pharm Tech Research*, Vol. 4(2), pp 643–647.

Vohra, K & V.K. Gupta, 2012, Pharmacognostic evaluation of *Lens culinaris* Medikus seeds, *Asian Pacific Journal of Tropical Biomedicine*, Vol. 2(3), pp 1221–1226.

Voutsina, N., A.C. Payne, R.D. Hancock, G.J. Clarkson, S.D. Rothwell, M.A. Chapman & G. Taylor, 2016, Characterization of the watercress (*Nasturtium officinale* R. Br.; Brassicaceae) transcriptome using RNASeq and identification of candidate genes for important phytonutrient traits linked to human health. doi: 10.1186/s12864-016-2704-4.

Wang, Y.F., J.J. Chen, Y. Yang, Y.T. Zheng, Sh.Z. Tang & Sh.D. Luo, 2002, New rotenoids from roots of mirabilis jalapa, *Helvetica Chimica Acta*, Vol. 85(8), pp 2342–2348.

Warwick, S.I., H.J. Beckie, A.G. Thomas & T. Mc Donald, 2000, The biology of Canadian weeds. 8. *Sinapis arvensis*. L.,Canadian *Journal* of *Plant* Science, Vol. 80(4), pp 939–961.

Weeden, N.F., 2018, Domestication of pea (*Pisum sativum* L.): The case of the abyssinian pea. doi: 10.3389/fpls.2018.00515.

Wenzig, E.M., U. Widowitz, O. Kuner … & R. Bauer, 2008, Phytochemical composition and in vitro pharmacological activity of two rose hip (*Rosa canina* L.) preparations, *Phytomedicine*, Vol. 15(10), pp 826–835.

Xiao, W., Sh. Li, S. Wang & Ch. Ho, 2016, Chemistry and bioactivity of *Gardenia jasminoides*, *Journal of Food and Drug Analysis*, Vol. 25(1), pp 43–61.

Yeyinou Loko, L.E., J. Toffa, A. Adjantin, A. JoelAkpo, A. Orobiyi & A. Dansi, 2018, Folk taxonomy and traditional uses of common bean (*Phaseolus vulgaris* L.) landraces by the sociolinguistic groups in the central region of the Republic of Benin, *Journal of Ethnobiology and Ethnomedicine*. doi:10.1186/s13002-018-0251-6.

Yilmaz, N., N. Yayli, G. Misir, … & N. Yayli, 2008, Chemical composition and antimicrobial activities of the essential oils of *Viburnum opulus*, *Viburnum lantana* and *Viburnum orientala*, *Asian Journal of Chemistry*, Vol. 20(50), pp 3324–3330.

Younesi, S., T. Nejadsattari & R. Hasani, 2015, Lichen floristic study from southeast of Ilam Province (Iran) (Dehloran, Abdanan, Darehshahr), *Biosciences Biotechnology Research Asia*, Vol. 12(2), pp 1373–1379.

Yu Ding, H., H. Ching Lin, C. Ming Teng & Y. Chang Wu, 2000, Phytochemical and pharmacological studies on Chinese paeonia species, *Journal of the Chinese Chemical Society*, Vol. 47(2), pp 381–388.

Zandpour, F., M.R. Vahabi, A.R. Allafchian & H.R. farhang, 2016, Phytochemical investigation of the essential oils from the leaf and stem of (Apiaceae) in Central Zagros, Iran, *Journal of Herbal Drugs*, Vol. 7(2), pp 109–116.

Zargari, A. 2014, *Medicinal Plants, First, Second, Third, Fourth and Fifth Volumes*, Tehran University Press, p.4274.

Zarger, M.S.S., F. Khatoon & N. Akhtar, 2014, Phytochemical investigation and growth inhibiting effects of *Salix alba* leaves against some pathogeni fungal isolates, *World Journal of Pharmacy and Pharmaceutical Sciences*, Vol. 3(10), pp 1320–1330.

Zheng Feei, M & H. Zhang, 2017, Phytochemical constituents, health benefits, and industrial applications of grape seeds, *Antioxidants (Basel)*, Vol. 6(3), pp 71–81.

Zorica, P., J. Bajic Ljubicic, R. Matic & S. Bojovic, 2017, First evidence and quantification of quercetin derivatives in dogberries (Cornus sanguinea L.), *Turkish Journal of Biochemistry*. doi:10.1515/tjb-2016-0175.

Zumrutdal, E & M. Ozaslan, 2012, A miracle plant for the herbal pharmacy; henna (*Lawsonia inermis*), *International Journal of Pharmacology*, Vol. 8, pp 483–489.

Index